VANISHING FISH

VANISHING FISH

Shifting Baselines and the Future of Global Fisheries

DANIEL PAULY

Foreword by JENNIFER JACQUET

DAVID SUZUKI INSTITUTE

GREYSTONE BOOKS
Vancouver/Berkeley

Copyright © 2019 by Daniel Pauly
Foreword copyright © 2019 by Jennifer Jacquet

19 20 21 22 23   5 4 3 2 1

All rights reserved. No part of this book may be reproduced, stored in a retrieval system or transmitted, in any form or by any means, without the prior written consent of the publisher or a license from The Canadian Copyright Licensing Agency (Access Copyright). For a copyright license, visit accesscopyright.ca or call toll free to 1-800-893-5777.

Greystone Books Ltd.
greystonebooks.com

David Suzuki Institute
davidsuzukiinstitute.org

Cataloguing data available from Library and Archives Canada
ISBN 978-1-77164-398-6 (pbk.)
ISBN 978-1-77164-399-3 (epub)

Editing by Nancy Flight
Copyediting by Rowena Rae
Cover design by Will Brown
Text design by Belle Wuthrich
Printed and bound in Canada on ancient-forest-friendly paper by Friesens

Greystone Books gratefully acknowledges the Musqueam, Squamish, and Tsleil-Waututh peoples on whose land our office is located.

Greystone Books thanks the Canada Council for the Arts, the British Columbia Arts Council, the Province of British Columbia through the Book Publishing Tax Credit, and the Government of Canada for supporting our publishing activities.

Canadä

CONTENTS

Foreword by Jennifer Jacquet   vii
Preface and Acknowledgments   xi

Duplicity and Ignorance in Fisheries   1
Aquacalypse Now: The End of Fish   21
Major Trends in Small-Scale Fisheries   33
ITQs: The Assumptions Behind a Meme   48
Putting Fisheries Management in Its Places   55
Fisheries Management: For Whom?   59
Fishing More and Catching Less   68
Bycatch Uses in Southeast Asia   73
On Reconstructing Catch Time Series   80
A Global, Community-Driven Catch Database   86
Catches Do Reflect Abundance   89
The Shifting Baseline Syndrome of Fisheries   94
Further Thoughts on Historical Observations   99
Consilience in Research   108
Focusing One's Microscope   116
Homo Sapiens: Cancer or Parasite?   121

Academics in Public Policy Debates   127
Worrying About Whales   132
Not the Fisheries Committee   137
My Personal Odyssey I: On Becoming a
Canadian Fisheries Scientist   145
My Personal Odyssey II: Toward a
Conservation Ethic for the Seas   156
My Personal Odyssey III: Having
to Science the Hell Out of It   169

Epilogue: *Some Gloom, but Surely No Doom*   196
Abbreviations and Glossary   198
Endnotes   213
Index   280

FOREWORD

MANY PEOPLE FIRST meet Daniel Pauly during one of his frequent speaking or film appearances, efforts he wryly refers to as his "fieldwork." My first impression of Daniel came through his writing. In 2003, when I was taking a course in marine ecology, I read his article "Fishing down the marine food web," published in *Science* in 1998, and with the first sentence, I became fascinated by the way Daniel thinks. The article begins: "Exploitation of the ocean for fish and marine invertebrates, both wholesome and valuable products, ought to be a prosperous sector, given that capture fisheries—in contrast to agriculture and aquaculture—reap harvests that did not need to be sown." The premise—that fish and invertebrates are products with which to make an economic sector prosperous—seems a little suspect to me now, and, as a pioneer in ecosystem-based approaches to fisheries, Daniel most likely would not write that line the same way today. But there is nonetheless something in his writing, as there is in all the essays in this book, that indicates Daniel has something to say and it matters to him how he says it.

Daniel's style can be partly attributed to the fact that, unlike most fisheries biologists, he has embraced both the humanities and the social sciences. He is especially fond of history, probably because of his tendency toward big-picture thinking. Daniel is also a man of many countries, and it is with some embarrassment for all of us native English speakers who have been closely edited by him that he spoke French and German well before he spoke English, which he wields with greater precision than we do. When he returned the first draft of the first paper he and I worked on together, he had written "good" on it but then crossed that word out and written "This will be good" below it. There is much to admire in the exactness of his observations and his writing.

Daniel no longer uses the term "harvest" to refer to fisheries catches, just as he has recently abandoned the term "stock" (see the preface). If we were to press him on this decision, he would have something clever to say, as he did when I once accused him of changing his mind about some issue, and, unfazed, he responded: "It's the best proof I still have one." We all shed our conditioning and become willing to see the world in new ways at different rates, and Daniel is faster than most. That is also the benefit that a collection like this affords—a chance to see how its writer has evolved over time. In these essays, Daniel refers to the threefold expansion of fisheries that he and his research team have identified and quantified: fishing farther offshore, fishing deeper, and fishing for new species. Similarly, the threefold expansion of Daniel's own thinking has fundamentally changed the fields of fisheries science and marine conservation, as well as the broader, public conversation.

First, Daniel expanded fisheries in space and time. He moved fisheries from a parochial endeavor in which every element is context- and species-specific and connected the dots to show us the benefits of considering fisheries as a global system. Because he had seen a lot of the world, Daniel was well positioned to think on a global scale, but he was also brave and tenacious enough to try to work that way too. I spent seven years with Daniel's Sea Around Us project and saw firsthand that the ability to do quality work on a global level requires not only intellect but also mettle, as well as serious endurance for drudgery. Daniel also shifted fisheries research from its all-too-convenient baselines of 1980 or 1970 and challenged researchers to think about fisheries over longer time frames. When his colleagues hosted a celebration at the University of British Columbia for Daniel's sixtieth birthday, ecologist Jeremy Jackson credited Daniel's single-page essay on shifting baselines (see "The Shifting Baseline Syndrome of Fisheries") with persuading him to pursue historical ecology (and as the impetus for Jackson et al.'s influential 2001 paper, published in *Science*, titled "Historical overfishing and the recent collapse of coastal ecosystems").[1]

Second, Daniel expanded fisheries science from a discipline dominated by industry concerns to one that considers multiple stakeholders. Instead of representing the interests of only the fishing industry, he works most closely with civil society groups. He pushed for the inclusion of small-scale fishers both in the global dataset and at fisheries negotiations. (At a talk where Daniel was asked why government subsidies favored industrial fisheries over small-scale fisheries, he responded: "Because small-scale fishermen don't play golf.") Through ecosystem modeling and other scientific work, he also fought to include marine animals in the conversation. He has, for example, repeatedly defended whales against the accusation by whaling country representatives that "they eat all of our fish" (and therefore that we should kill them to increase global fisheries catches) with scientific evidence demonstrating that the distribution and consumption patterns of whales do not significantly overlap with the operating areas of, and the species exploited by, fisheries.

Third, Daniel has pushed—to a lesser extent, but one that should not be overlooked—for fish and invertebrates not to be seen exclusively as seafood or commodities. He has asked that, as with whales, we consider fish not just as species but also as individual animals. Throughout his career Daniel has increasingly veered from the views held by many of his scientific colleagues who endorse a status quo—one that Daniel, as scientist and as citizen, thinks is pernicious both for his discipline and for society at large. May we all aspire to this kind of critical thinking and courage.

On a final note, in Western society, particularly in the United States, science and scientists bear most of the burden of proof regarding the potential for negative impacts of new developments. Although, in theory, we should operate under a precautionary principle, in practice, scientists must demonstrate harm to the ecosystem, but extractive industries need not demonstrate their lack of negative impacts. Daniel has repeatedly shouldered this burden for all who believe that the interests of civil society, small-scale fishers, and the animals in the ocean should also be considered. This book is a tribute to his willing and adept work on behalf of this disparate group.

JENNIFER JACQUET
NEW YORK UNIVERSITY, 2019

PREFACE AND ACKNOWLEDGMENTS

BOTH FISHERIES SCIENTISTS and the general public often perceive fisheries science as the study of localized activities—the catching of specific freshwater or marine fish at specific places (a stretch of a river, a bay, or a sea), using a specific type of gear. However, fisheries nowadays, especially marine fisheries, are all part of an expanding, global system, interconnected by powerful market forces and the range of modern distant-water fleets, which operate everywhere there are fish, undeterred by distance, water depth, storms, or ice cover. Many fisheries can now even operate where fish have ceased to be abundant, as they rely on government subsidies, which, along with the increasing use of semi-enslaved crew, enable them to eke out a profit from depleted fish populations.

Except for the epilogue, written in January 2017, and two new essays also written in 2017, this book consists of previously issued, single-authored essays of which the earliest—on shifting baselines—was published in 1995, and all the others between 1996 and 2016. As some of their topics overlap, readers are free to skip repetitious paragraphs. The text of the original essays has not been updated except for the occasional removal of single words such as "recently" and the deletion of a few paragraphs dealing with issues of no present interest. The book has been made more current, however, via the addition of detailed endnotes (identified as N.N. for new notes) linking the ideas proposed in its earlier essays with references to current information and debates. One exception is the word "stock," which I previously held to be roughly synonymous with "exploited fish population," but which I now realize is part of the technocratic ideology that isolates us from Nature. I have thus deleted or replaced this word whenever it could be done without loss of understanding.

Also included, in spite of unavoidable overlap due to their subject matter, are three decidedly autobiographical essays, because the paths that my life and career took were weird enough to be of interest to the reader. Also, I was a very observant, if horrified, witness to the depletion of marine life that occurred in the later part of the 20th century, and so I have something to say about that.

Most of the contributions presented here were assembled in the fall of 2016, while I was on a mini-sabbatical at the Department of Environmental Studies of New York University, whose members, notably Drs. Dale Jamieson and Jennifer Jacquet, I thank for making me feel welcome. In addition, I thank Jennifer for her generous foreword. I also thank Sandra Wade Pauly for converting my disparate PDFS and other files into a coherent whole and standardizing their references to endnotes, work without which this collection of essays would not have seen the light of day.

Finally, I thank the Natural Sciences and Engineering Research Council of Canada for support in the late 1990s, and I thank the Pew Charitable Trusts, notably Ms. Rebecca Rimel and Dr. Joshua Reichert, and the Paul G. Allen Family Foundation for their support of the Sea Around Us project from mid-1999 to mid-2014 and mid-2014 to mid-2017, respectively.

DUPLICITY AND IGNORANCE IN FISHERIES[2]

SETTING THE STAGE

WHAT IS THE CATCH?

THIS COLLECTION OF essays deals mainly with marine fisheries, which are one of the three major ways we use, and thus interact with, the oceans. (The other two uses are for transporting goods and getting rid of our trash and excrement.)

The impact of fisheries on ocean ecosystems can be measured in various ways. The simplest way is to estimate the quantities of life we extract from them by monitoring trends in the catches in different parts of the world. This seems obvious, but strangely, some fisheries scientists disagree; this topic is discussed in this essay and in several others.

Statistics covering the "visible" part of global fisheries, that is, official fisheries catch data, have existed since the 1930s,[3] when the unfortunate League of Nations first attempted to report on the world's economy. The United Nations (UN), founded in 1945, continued this effort[4] in 1950, when the Food and Agriculture Organization of the United Nations (FAO) began to issue its annual *Yearbook of Fisheries*

Statistics. The annually revised and updated data in these yearbooks—now a database available online (at http://www.fao.org/)—are widely used by the FAO and other UN agencies but also by academics and other researchers to track the development of fisheries by country and region, as well as globally, and to pronounce on their future prospects.[5]

However, many of these researchers are unaware of the manner in which this dataset is created,[6] and of its deficiencies (notably a huge "invisible" catch). These deficiencies will have to be addressed (especially because, as the phrase goes, this is "the only dataset we have") if we want to seriously address the overexploitation of marine ecosystems.[7]

In the first few decades after World War II, the growth of marine fisheries was very rapid, whether it is measured by input into the fisheries (invested capital, cumulative vessel tonnage, engine power,[8] etc.) or output (tonnage caught or ex-vessel value of the landings—the price received at the point of landing[9]). This period, which created the basis for the worldwide industrialization of fisheries, was also a time when fisheries appeared to behave like any other sector of the economy, with increased inputs leading to increased outputs. This is the rationale behind the subsidization of fisheries, a subject to which we will return.

THE TOXIC TRIAD OF FISHERIES

THIS PERIOD WAS also one of massive fisheries collapses, in which the abundance that sustained entire fishing fleets, processing plants, and thousands of workers and their families disappeared nearly overnight.[10] The California sardine fishery was one of these, although it does live on in John Steinbeck's *Cannery Row*. Others, more prosaically, were rebuilt after a few years: for example, the fishery for Atlanto-Scandian herring[11] and the Peruvian anchoveta fishery, whose first massive collapse occurred in.1972.[12] The Peruvian example best illustrates an approach already prevalent in the heyday of the California sardine fishery: blame the environment. Thus, in Peru it was El Niño that "did it," never mind the fact that the catch the year before the collapse was over 16 million metric tons[13,14] rather than the officially reported 12 million metric tons, which itself exceeded what the best experts of the time (John Gulland, Bill Ricker, and Garth Murphy) had recommended as sustainable.

Various concepts have been deployed to understand these events. One of these is the "tragedy of the commons,"[15] which can be made to explain why the pathologies mentioned above were likely to occur in the largely unregulated fisheries then prevalent. The concept proposed here is that of a "toxic triad" of (1) underreporting the catch, (2) overfishing (i.e., ignoring the scientific advice available at the time), and (3) blaming "the environment" for the ensuing mess. This concept could be easily extended to cover more pathological aspects of fisheries (thus leading to a toxic tetrad, etc.), but the three elements mentioned here are sufficient for our purposes.

The toxic triad existed long before its effects became widespread. When they did, however, a battery of new terms had to be coined to deal, at least conceptually, with the new development. Hence the coining of the word "by-catch" by W.H.L. Allsopp[16] and the emergence of the concept of IUU (illegal, unreported, and unregulated fisheries), without which the stark reality these terms describe cannot be fully apprehended.

The toxic triad was firmly in place when, in 1975, catches peaked in the North Atlantic, before going into a slow decline that continues to this day.[17] However, this decline became unmistakable when the giant population of northern cod off Newfoundland and Labrador collapsed, bankrupting an entire Canadian province and setting off a frantic search for something to blame (hungry seals, cold water, etc.) other than the out-of-control fishing industry.[18] We will return to this event repeatedly.

A THREEFOLD EXPANSION

THE TOXIC TRIAD provided a rationale for expansion, which occurred in three dimensions:

GEOGRAPHIC EXPANSION

The relatively well-documented freshwater and coastal fisheries of ancient times had the capacity to induce severe decline in, and even extirpate, vulnerable species of marine mammals, fish, and invertebrates, as documented by a variety of sources.[19] However, it is only since the onset of industrial fishing, using vessels powered by fossil fuel (a development that occurred in the 1880s, when the first steam-powered trawlers were deployed), that successive depletion of inshore fish populations, followed by that of more offshore populations, has become routine.[20] Thus, in the North Sea, it took only a few years for the accumulated biomass of coastal flatfish and other groups to be depleted and for the trawlers to be forced to move on to the central North Sea, then farther, all the way to Iceland.[21]

A southward expansion soon followed, toward the tropics,[22] leading to the development of industrial fishing in the nascent "Third World," often through joint ventures with European (especially Spanish) or Japanese firms. This expansion created new conflicts over access to resources and intensified earlier ones—hence the protracted "cod war" between Iceland and Britain from 1958 to 1975 and the brief "turbot war" of March 1995 between Canada and Spain.[23] At the close of the 20th century, the demersal resources of all large shelves of the world, all the way south to Patagonia and Antarctica,[24] had been depleted, mainly by trawling, along with those of seamounts and oceanic plateaus.[25]

From 1950 to 1980, industrial fisheries expanded their reach by about 0.4 million square miles per year;[26] the expansion rate increased to 1.1 to 1.5 million square miles per year in the 1980s and then declined, while the southward expansion proceeded at the remarkably constant rate of 0.8 degree of latitude per year from the 1950s onward.[27] By 2000, the geographic expansion was essentially over, and what remained were the two forms of expansion described below.

BATHYMETRIC EXPANSION

The second dimension of the expansion of fishing was at depth (i.e., offshore), which affected both the pelagic and demersal realms. In the pelagic realm, the exploitation of tuna, billfishes, and increasingly sharks (for their fins[28]) by longlines and similar gear has strongly modified oceanic ecosystems, which now have much reduced biomass of large predators.[29] This effect is intensified by the use of fish-aggregating devices (FADS), which, starting around the Philippines,[30] have spread throughout the inter-tropical belt and have made small tuna and other fish that could not be captured before accessible to fisheries, thus constituting an additional expansion of sorts.

In the demersal realm, trawlers were deployed that can and increasingly do reach depths of a mile or more,[31] yielding a catch increasingly dominated by slow-growing, deep-water species with low productivity, which cannot be exploited sustainably.[32,33,34] Therefore, given that the high seas (the waters outside Exclusive Economic Zones or EEZs; see below) are legally unprotected, their oceanic plateaus and seamounts are subjected to intense localized fishing pressure, leading to the collapse of the resources; this process is then repeated on the adjacent plateau or seamount.[35] This fishing mode is no more sustainable than tropical deforestation.

The resulting changes in biomass induce—notably via altered food webs—massive changes in demersal and pelagic communities; these changes can be demonstrated and quantified in various ways.[36] The marine trophic index (MTI)—the mean trophic level of fisheries landings[37]—is one of the most widely used indicators for this purpose. The MTI is declining throughout the world,[38] meaning that, increasingly, fisheries catches are based on small fish and the invertebrates at the base of the ocean's food webs.[39]

TAXONOMIC EXPANSION

The term "taxonomic expansion" refers to the catching and processing of previously spurned taxa, such as the ugly monkfish or jellyfish.[40] This form of expansion, which intensifies the effect of geographic and bathymetric expansion, is the reason that markets in the United States, Canada, and Europe increasingly display unfamiliar seafood, offering many opportunities for mislabeling products and misleading consumers[41,42,43]—one of the reasons for the word "duplicity" in the title of this essay.

DIGRESSION 1:
EXCLUSIVE ECONOMIC ZONES

IN THE EARLY 1980s, the decade-long deliberations that had been triggered when various maritime countries unilaterally declared ownership of huge swaths of their coastal waters led to the United Nations Convention on the Law of the Sea (UNCLOS). Because of the UNCLOS, all maritime countries could claim EEZs of up to 200 nautical miles and thus (if they had the political clout) could throw out the distant-water fleets that were then operating wherever they pleased.[44] Some more powerful countries did throw out the distant-water fleets that had operated along their coasts but then began subsidizing the development of national fleets that soon became as destructive as the foreign fleets had been.[45] In the United States and Canada, this development eventually led to the collapse of cod in New England and along the Canadian East Coast. Other countries—notably, several in Northwest Africa—tried to expel the distant-water fleets operating in their waters. However, without political clout, they were susceptible to blackmail (in cases where their negotiators were honest) or bribery (in cases where they weren't). As a result, European and East Asian distant-water fleets are still operating in that region.[46,47]

The continued presence of distant-water fleets from countries of the European Union is based on access "agreements," most of which are triumphs of raw political power over the rhetoric of partnership and development aid—another realm in which duplicity reigns supreme. As for distant-water fleets from East Asia, the rhetoric is different. China has no rhetorical stance: its fisheries operate mostly on what appear to be private agreements with local politicians that stay under everybody's radar,[48] showing up in the press only where its trawlers get into conflicts with local fishers.[49] This is a far cry from the situation of just a few years ago, when China was over-reporting its catches (see below).

Japan, in contrast, manages to add insult to injury: its fisheries experts and embassies in Northwest Africa argue that it is the whales that are responsible for the decline of the fish populations and that therefore the countries in question should help re-establish "ecosystem balance" by, among other things, voting at the International Whaling Commission in support of Japan killing more whales.[50] This line of argument, which would be specious anywhere, is particularly duplicitous in Northwest Africa, where distant-water fleets and overgrown "small-scale" fisheries have been unequivocally shown to be the cause of widespread declines in fish populations[51] and where baleen whales occur mainly during the reproductive season when they do not feed (see the essay titled "Worrying About Whales").

Strangely enough—and this is the reason for "ignorance" in the title of this essay—this line of argument was successful in that, probably with the help of other, more practical enticements, it has redirected the scarce research resources of several countries in Northwest Africa toward conducting costly "whale surveys"—this in a region that usually does not have observers on board the vessels of distant-water fleets and that, in fact, has no practical way of even estimating their catches. A similar misdirection of emphasis has occurred among smaller Caribbean and South Pacific states, for the same reasons, and it is equally pernicious.[52,53]

THE CRISIS OF FISHERIES

DIRECT AND INDIRECT DRIVERS

THE EXPANSION TRENDS established in the 1980s and 1990s have led to the crisis at the onset of the 21st century, of which the following are major elements.

First, there is, in the global fisheries sector, an excessive number of fishing vessels, variously estimated as two to three times what is required to generate present catches.[54,55,56] This is probably an underestimate, given that the increased efficiency of vessels in locating and catching fish ranges from 2 to 3 percent a year across a wide range of vessel types, implying that the effective fishing effort (i.e., their ability to catch fish) doubles every fifteen years.[57]

Second, the biomass of traditionally targeted large fish (cod and other demersal fish, tuna and other large pelagic fish) has been reduced by at least one order of magnitude since the onset of industrial exploitation.[58,59,60] The generality of these findings has been contested (see below), but they can be straightforwardly reproduced by anyone willing to reconstruct population sizes before industrial exploitation, as was done for New England cod,[61] or the demersal biomass around the United Kingdom.[62] Without such reconstructions, arguments about depletion will essentially be useless, as subjective perceptions of abundance are biased by shifting baselines (see also the essay titled "The Shifting Baseline Syndrome of Fisheries"). This bias has been empirically shown to be extremely strong.[63]

Finally, one aspect of global fisheries—but one often not perceived as the scandal it is—is that about one-quarter of the world's industrial catch (mainly sardines, anchovies, mackerel, and other small pelagic fishes) is wasted as animal feed (mainly as fishmeal, of which about half is consumed by aquaculture, including mariculture), although it could easily be converted to human food.[64,65] As such, it would contribute far more to human nutrition (including the supply of omega-3 fatty acids) than does aquaculture (which inserts a trophic step in the food chain between these fish and humans), while avoiding the bioaccumulation of persistent organic pollutants, which makes farmed carnivores such as salmon so problematic.[66]

Note that even if the supply of small pelagics may have increased (because of predator depletion[67]), the expansion of aquaculture is still going to be limited, at least if aquaculture is defined as the raising of carnivorous fish (e.g., salmon, sea bass, and tuna), as is usually implied in Western countries and regions. In the Mediterranean, for example, the farming of high trophic level fish—or "farming up"[68]— has expanded. There, large quantities of small pelagics are fished to feed relatively few farmed fish (mainly tuna), leaving no food for marine mammals[69] and fewer fish for people who cannot afford to eat bluefin tuna sushi. In contrast, in China, where almost two-thirds of aquaculture production is reportedly occurring, the major farmed species are herbivorous freshwater fish and marine bivalves, neither of which requires fishmeal.[70]

Over 50 percent of the fish caught in the world is traded internationally, and many industrialized countries either have huge distant-water fleets (as Spain still does) or purchase most of the fish they consume (as, for example,

Germany and Japan do). In any case, there is a large net flow of fish from developing to industrialized countries, with serious consequences for the food security of the protein-deficient, least-developed countries.[71,72,73]

Various market-based initiatives in industrialized countries are centered on the belief that by changing consumer behavior they can change the way fish are exploited.[74,75] The UK-based Marine Stewardship Council (MSC) is the best known of these, along with the multitude of seafood guides that, as their name implies, are meant to advise consumers as to the "sustainability" of the species offered in seafood markets and restaurants.[76,77] However, even if this goal were reached, it would still not solve the food security problem caused by the transfer of fish from developing to industrialized countries.

On top of it all, government subsidies to fisheries—the grease that keeps the whole creaky system going—are currently estimated at US$25 to $35 billion per year,[78,79] up from the previously accepted figure of $14 to $20 billion per year.[80] Of this total, over half are capacity-enhancing, or "bad," subsidies. This term applies especially to fuel subsidies, which allow profitable exploitation of depleted fish populations and thus directly contribute to the problems described above. However, these problems could be resolved—through the World Trade Organization,[81] for example—as most industrial fisheries, and particularly the fuel-intensive trawl fisheries, now depend on subsidies, especially fuel subsidies, which make them extremely sensitive to changes in the cost of fuel.[82]

SUBJECTIVE FACTORS AND MASKING EFFECTS

IN ADDITION TO the objective factors or drivers mentioned above, there are a number of subjective elements, some bordering on duplicity (and some crossing that border), that contributed to the crisis being masked, or at least misunderstood, and thereby contributed to the decline of marine biodiversity and overexploitation of the ecosystem.

The first of these factors was the massive over-reporting of catches by China through the 1990s, which misled the FAO and the world into believing that global landings were increasing, whereas they were, in fact, slowly decreasing.[83] This over-reporting occurred because an independent statistical system does not exist in China, meaning that favorable production statistics can be manufactured by mid-level officials, including those in the fisheries sector, seeking advancement.[84,85] The FAO, which now presents world fisheries statistics with and without China, has doubts about Chinese aquaculture statistics as well.[86]

Another masking factor is that the *per capita* consumption of fish in industrialized countries, especially in the European Union and the United States, is still increasing. Given declining global catches,[87] one can assume that *per capita* consumption in developing countries (excluding China) is declining. Reliable data on fish consumption in developing countries do not exist to test this assumption (which will be convenient to some). In the meantime, consumers in the European Union and the United States are left to enjoy *frissons* of guilt when ordering seafood not sanctioned by their many seafood guides.

However, the most potent masking factors—because they provide governments with the excuses they need not to intervene and counter negative trends—are, as in the case of global warming, the denials of self-styled "skeptics" and their misuse of "uncertainty."[88] The skeptics are effective because science needs skepticism and must recognize uncertainty, and fisheries science is no exception.

In a brilliant paper now repudiated by its second author, Don Ludwig, Ray Hilborn, and Carl Walters[89] outlined how scientific uncertainty is being used in fisheries to hold off intervention until it is too late to prevent the collapses of exploited fish populations; that is, scientific uncertainty is not used in a precautionary fashion. This problem can be aggravated when the skeptics combine their denials with innuendos about the objectivity and ethics of conservation-oriented scientists, the journals that publish their research, and the donors that fund it. Thorough refutations, as occurred in response to Bjorn Lomborg's denials of global warming,[90] are still pending in the case of fisheries. They still have to be written, lest the Panglossian views espoused by the fishy "merchants of doubt"[91] manage to divert us from describing and then repairing what is wrong with our fisheries.

DIGRESSION II:
CATCH UNDER-REPORTING

EXCEPT FOR CHINA, whose political system encourages the over-reporting of domestic catches, and a few strongmen insisting on increasing catches in the countries they thought they controlled (e.g., Ferdinand Marcos in the Philippines in the early 1980s), the catch data available to the public and most scientists are biased downward and against small-scale fisheries. This occurs in two steps: (1) government scientists generally study—and the statistical system they set up usually monitors—only commercial fisheries (and not recreational and small-scale, artisanal, or subsistence fisheries, even if they collectively land the bulk of the national catch[92,93]), and (2) the national agencies that report national catches to the FAO, which compiles and maintains the only global database of fisheries statistics, are usually not the departments of fisheries or similar entities but the ministries of agriculture or finance or their statistical offices. These agencies all tend to emphasize "cash crops," meaning exportable products such as shrimp and tuna, while giving short shrift to—and at worst completely ignoring—the catches of small-scale fisheries, even though it is these catches that feed their rural populations.[94,95]

These two problems are so widespread that in the mid-2000s the Sea Around Us[96] initiated a systematic reconstruction of the real catch (i.e., including those of illegal and/or unreported fisheries) of all maritime countries of the world. This reconstruction was completed in early 2016 and the main finding was that "global marine fisheries catches are higher than reported and declining,"[97] a topic to which we shall return.[98]

OPPORTUNITIES FOR RENEWAL

THE RENEWAL OF FISHERIES SCIENCE

CLEARLY, WE NOW have a situation in which a substantial portion of the fishing industry is willing to sacrifice its own future, a future that can be sustainable only if the resources these fisheries exploit are allowed to recover and to rebuild their biomass. The most important task for a renewal of fisheries and fisheries research is therefore the reduction of the overall fishing effort. Without this, nothing else will work. Ecosystem-based considerations will also play a part,[99,100] and will mean ensuring, among other things, that we do not attempt to maximize catches of either predators or their prey. No-take marine reserves will have to be perceived not as scattered small concessions to conservationist pressure but as a legitimate and obvious management tool designed to re-establish natural refuges lost to the geographic and bathymetric expansions described above.[101]

A major goal for future management regimes will be to avoid the extinction of species previously protected by their inaccessibility to fishing gear, as well as to account for the effects of global warming.[102,103,104] This goal would link fisheries scientists with the vibrant communities of researchers now working on marine biodiversity and conservation issues. Such a linking is not easily brought about, however, as suggested in the following section.

CHANGES FOR WHICH THE TIME HAD COME

ONE OF THE few good things about getting old is that one develops a fine-grained appreciation of the various forms of change. One such form is the insidious gradual shifting of the baselines we use to evaluate the state of the biodiversity surrounding us.[105] Another type of change is that which occurs when insights pile up in society, creating a tension that is not released before some abrupt event or "tipping point" occurs.[106] Examples of such events that I have witnessed include the May 1968 Parisian revolt against an increasingly autocratic Charles de Gaulle and the 1986 People Power revolution in Manila against the dictator Ferdinand Marcos. Other such changes were the civil rights movement in the United States, of which I saw only the tail end and which heralded the emergence of a new mindset that could no longer comprehend how the old one was ever acceptable.

Change also occurred in my chosen profession, fisheries science. When I was a student, I was taught that the work I was learning to do was supposed to be used by fisheries managers to ensure that fisheries resources were used as efficiently and effectively as possible and their exploitation put on some rational footing so that society as a whole would benefit. I worked first in developing countries, many of them with mighty fisheries, but the fisheries scientists whom I interacted with, and for whom I adapted some of the classical "stock assessment" models for use in the tropics,[107] were in no way connected to those making decisions about fisheries. Specifically, they had no connection to the heads of fishing enterprises (often, if misleadingly, called fishermen) or the financiers of fishing ventures, not to mention the politicians who facilitate and subsidize these projects. In other words, these fisheries scientists had no way of effecting change based on scientific evidence.

Later, when examining fisheries in Europe and North America, as well as globally, I found that this lack of access was the rule and that well-managed fisheries were the exception. I could also see the damage that fisheries do to marine ecosystems and biodiversity, and fisheries biologists' lack of a conceptual apparatus for dealing with biodiversity-related issues. In fact, they were not even considered legitimate research issues. Therefore, at several forums—notably, the ICES Annual Meeting in Bruges, Belgium, in 2000, and the 4th World Fisheries Congress in Vancouver, Canada, in 2004—as the keynote speaker, I tried to argue from within the fisheries profession for the need to extend our discipline from one implicitly interested in keeping fishing fleets operating to a broader one, devoted to maintaining marine ecosystems and the wildlife embedded therein, upon which the fisheries ultimately depend. I did this because I considered the argument made by ecologists who viewed ecosystems as more than larders[108] from which we extract what we want—without accounting for the externalities of fishing—to be legitimate.

For a while, I believed that these efforts, paralleled by those of numerous other researchers—notably Jeremy Jackson and his colleagues[109] and the late Ransom A. Myers and his co-workers—would be successful in generating a new consensus. Who can argue against the need to maintain the fish that maintain the fishery? But now, with the publication of a host of contrarian articles and their apparently positive reception by a number of my colleagues, I see that this is not obvious at all.

Instead, what has emerged in our discipline of fisheries science is a discussion about standards of evidence and even about what constitutes evidence. Such "meta-discussion," or discussion about the ways we ought to conduct ourselves, is indicative of a deep malaise and of the fundamental changes that have occurred, where two schools now fight for supremacy, to represent the discipline as a whole.

One school centers on the profitability of fishing enterprises and on fishing "rights" (see below), the other on marine ecosystems and their ability to generate services, including fisheries catches. Moreover, between them, name-calling has become the fashion in a way that would have been impossible in the times of Ray Beverton and, I presume, the other gentlemen (yes, they were all men) who founded quantitative fishery science.

I am not a neutral observer in this—far from it; thus, I think it is obvious that the next generations will expect fisheries to perform well not only in operational and financial terms but also in ecological terms. This will involve overcoming the results of the conceptual sleight of hand by a number of fisheries economists, in which the need for fishers to have predictable access to the resources[110] has been turned into claims that these resources should be handed over to fleet owners in perpetuity, along with exclusive "rights." It is ironic that such privatization of a public good should be called rights-based fishing.[111]

In most countries, the fisheries resources in the EEZs belong to the state (i.e., to us all), and these resources could be managed in the same manner as public forests or rangelands, through leases or temporary licenses that can be auctioned.[112,113] This would help address the overcapacity problem as efficiently as through "rights-based fishing" while avoiding the privatization of a public good, of which we should be wary, given the experience of the last decades, culminating in a financial crisis largely caused by the unmooring of the market from all ethical constraints.

BUT THERE ARE ALTERNATIVES

THERE ARE BASICALLY two alternatives for the future of fisheries science and management. One is to continue with business as usual, by which I mean accommodating subsidy-driven overcapacity without bothering about externalities—that is, damage to the ecosystems. This would lead not only to further depletion of biodiversity but also to intensification of "fishing down the marine food web," which means that ultimately marine ecosystems would be transformed into dead zones.[114] The other alternative is to convert fisheries science and management into life-affirming disciplines, which, rather than maximizing return to fisheries, would be devoted to implementing some form of ecosystem-based fisheries management, requiring consideration of more stakeholders than the fishing industry alone. This transformation would also require extensive use of ocean zoning and spatial closures (permanent or seasonal bans on fishing in defined areas), including permanent no-take marine protected areas or marine reserves.[115]

Marine reserves must be at the core of any scheme intending to put fisheries on an ecologically sustainable basis. Such reserves currently cover a very small fraction of the world's oceans, but the 5 percent annual increase of the cumulative area protected that prevailed through the 1980s and 1990s[116] has jumped up in the 21st century as a few massive marine reserves have been created in the Pacific and Indian Oceans.[117] Thus, although the 10 percent coverage in 2010 agreed to by the countries that are members of the Convention on Biological Diversity[118] has not been reached, the idea has now caught on and future targets may be attained. This is heartening, because if marine biodiversity is to be maintained and functional ecosystems re-established where uncontrolled exploitation has obliterated them, we will have to set up larger marine reserves, at a faster pace, as is also advocated by most marine ecologists and by all non-governmental organizations working on the marine environment.[119]

It is comforting that Ramon Margalef anticipated this thought[120] long before it became fashionable when he wrote: "Probably the best solution would be a balanced mosaic, or rather a honeycomb, of exploited and protected areas. Conservation is also important from the practical point of view: lost genotypes are irretrievable treasures, and natural ecosystems are necessary as references in the study of exploited ecosystems."

AQUACALYPSE NOW: THE END OF FISH[121]

A VAST PONZI SCHEME[122]

OUR OCEANS HAVE been the victims of a giant Ponzi scheme, waged with Bernie Madoff–like callousness by the world's fisheries. Beginning in the 1950s, as their operations became increasingly industrialized, with onboard refrigeration, acoustic fish-finders, and, later Geographic Positioning Systems, or GPS, the fishing fleets first depleted populations of cod, hake, flounder, sole, and halibut in the Northern Hemisphere. As the abundance of those fish declined, the fleets moved southward, to the coasts of developing countries, and, ultimately, all the way to the shores of Antarctica, searching for icefishes and rock cods, and finally for the small, shrimplike krill.

As the bounty of coastal waters dropped, fisheries moved farther offshore, to deeper waters. And, finally, as the larger fish began to disappear, boats began to catch smaller, uglier fish that had never before been considered fit for human consumption. Many were renamed so that they could be more easily marketed.[123] The suspicious slimehead[124] became the delicious orange roughy, while the worrisome Patagonian toothfish[125] became the wholesome Chilean sea bass. Others, like the homely hoki,[126] were cut up so that they could be sold sight unseen as fish sticks and filets in fast-food restaurants and the frozen-food aisle.

The scheme was carried out by nothing less than a fishing-industrial complex—an alliance of corporate fishing fleets, lobbyists, parliamentary representatives, and fisheries economists. By hiding behind the romantic image of the small-scale, independent fisher, they secured political influence and government subsidies far in excess of what would be expected, given their minuscule contribution to the GDP of advanced economies—in the United States, even less than that of the hair salon industry. In Japan today, huge, vertically integrated conglomerates, such as Taiyo or the better-known Mitsubishi, lobby their friends in the Japanese Fisheries Agency and the Ministry of Foreign Affairs to help them gain access to the few remaining plentiful tuna populations, such as those in the waters surrounding South Pacific countries. Beginning in the early 1980s, the United States, which had not traditionally been much of a fishing country, began heavily subsidizing U.S. fleets,[127] producing its own fishing-industrial complex, dominated by large processors and retail chains.

Today, the world's governments provide over US$30 billion in subsidies each year[128]—about one-third of the value of the global catch—that keep fisheries going, even when they have overexploited their resource base. As a result, there are between two and four times as many boats as the annual catch requires; yet the funds to "build capacity" keep coming.

The jig, however, is nearly up. In 1950, the Food and Agriculture Organization of the United Nations (FAO) estimated that, globally, we were catching about 17 million metric tons annually of fish (cod, mackerel, tuna, etc.) and invertebrates (lobster, squid, clams, etc.). Reported marine catches peaked at about 90 million metric tons per year in the mid-1990s, and they have been declining since.[129] Much like Madoff's infamous operation, which required a constant influx of new investments to generate "revenue" for past investors, the global fishing-industrial complex has required a constant influx of new "stocks" to continue operating. Instead of restricting its catches so that fish can reproduce and maintain their populations, the industry has simply fished until fish populations were depleted and then moved on to new or deeper waters and to smaller and stranger fish. And, just as a Ponzi scheme will collapse once the pool of potential investors has been drained, so too will the fishing industry collapse as the oceans are drained of life.

OUR IMPACTS ON THE OCEAN

UNFORTUNATELY, IT IS not just the future of the fishing industry that is at stake but also the continued health of the world's largest ecosystem. While the climate crisis regularly gathers front-page attention, people—even those who profess great environmental consciousness—continue to eat fish as if fishing were a sustainable practice. But eating a tuna roll at a sushi restaurant should be considered no more environmentally benign than driving a Hummer or harpooning a manatee. In the past fifty years, we have reduced the populations of large commercial fish, such as bluefin tuna, Atlantic cod, and other favorites, by a staggering 90 percent.

One study, published in the prestigious journal *Science*, forecast that by 2048 all commercial fish populations will have "collapsed," meaning that they will be generating 10 percent or less of their peak catches.[130] Whether or not the particular year, or even decade, of the peak catch is correct, one thing is clear: fish are in dire peril, and, if they are, then so are we.

The extent of the fisheries' Ponzi scheme eluded government scientists for many years. They had long studied the health of fish populations, of course, but typically laboratories would focus only on the species in their country's waters. Moreover, those studying a particular species in one country would communicate only with those studying that same species in another. Thus, they failed to notice an important pattern: popular species were sequentially replacing each other in the catches that fisheries were reporting, and, when a species faded, scientific attention shifted to the replacement species. At any given moment, scientists might acknowledge that one-half or two-thirds of fisheries were being overfished, but when the abundance of a particular fish had strongly declined, it was simply removed from the denominator of the fraction. For example, the Hudson River sturgeon was not counted as overfished once it disappeared from New York waters; it simply became an anecdote in the historical record.[131] The baselines just kept shifting,[132] allowing us to continue blithely damaging marine ecosystems.

It was not until the 1990s that a series of high-profile scientific papers demonstrated that we needed to study, and mitigate, fish depletions at the global level. These studies showed that phenomena previously observed at local levels—for example, the disappearance of large species from fisheries' catches and their replacement by smaller species—were also occurring globally. It was a realization akin to understanding that financial meltdowns are due not to the failure of a single bank but rather to the failure of the entire banking system. And it drew a lot of controversy.

TWO TYPES OF SCIENTIFIC RESPONSES

THE NOTION THAT fish are globally imperiled has been challenged in many ways—perhaps most notably by fisheries biologists, who have questioned the facts, the tone, and even the integrity of those making such allegations. Marine ecologists are concerned mainly with threats to the diversity of the ecosystems they study, and so, they frequently work in concert with environmental nongovernmental organizations (NGOs) and are often funded by philanthropic foundations.[133] In contrast, fisheries biologists traditionally work for government agencies, like the National Marine Fisheries Service within the U.S. Department of Commerce, or as consultants to the fishing industry, and their chief goal is to protect fisheries and the fishers they employ. I myself was trained as a fisheries biologist in Germany, and, while they would dispute this, the agencies for which many of my former classmates work clearly have been captured by the industry they are supposed to regulate. Thus, there are fisheries scientists who, for example, write that cod have "recovered" or even "doubled" their numbers when, in fact, they have increased merely from 1 percent to 2 percent of their original abundance in the 1950s or earlier.

Yet despite their different interests and priorities—and despite their disagreements about the "end of fish"—marine ecologists and fisheries scientists both want there to be more fish in the oceans. This is partly because both are scientists, who are expected to concede when confronted with strong evidence. And, as with global warming, the evidence is overwhelming: fish populations are declining in most parts of the world. Ultimately, the important rift is not between these two groups of scientists but between the public, which owns the sea's resources, and the fishing-industrial complex, which needs fresh capital for its Ponzi scheme. The difficulty lies in forcing the fishing-industrial complex to catch fewer fish so that populations can rebuild.

It is essential that we do so as quickly as possible because the consequences of an end to fish are frightening. To some Western countries, an end to fish might simply seem like a culinary catastrophe, but for half a billion people in developing countries, particularly in Africa and South and Southeast Asia, fish are the main source of animal protein. What's more, fisheries are a major source of livelihood for hundreds of millions of people. A World Bank report found that the income of the world's 30 million small-scale fishers is shrinking.[134] The decrease in catch has also dealt a blow to a prime source of foreign-exchange earnings, which impoverished countries—ranging from Senegal in West Africa to the Solomon Islands in the South Pacific—rely on to support their imports of staples such as rice.

THE OCEANS' RESPONSES

THE END OF fish would disrupt marine ecosystems to an extent that we are only now beginning to appreciate. Thus, the removal of small fish in the Mediterranean to fatten bluefin tuna in pens is causing the "common" dolphin to become exceedingly rare in some areas, and local extinction is probable. Other marine mammals and seabirds are similarly affected in various parts of the world. Moreover, the removal of top predators from marine ecosystems has cascading effects, leading to an increase in jellyfish and other gelatinous zooplankton[135] and to the gradual erosion of the food web within which fish populations are embedded. This is what happened at the turn of this century off the coast of southwestern Africa, where an upwelling ecosystem similar to that off California, previously dominated by fish such as hake and sardines, has been taken over by millions of tons of jellyfish.[136]

Jellyfish population outbursts are also becoming more frequent in the northern Gulf of Mexico, where the fertilizer-laden runoff from the Mississippi River fuels uncontrolled algal blooms. The dead algae then fall to a sea bottom from which shrimp trawling has raked all animals capable of feeding on them, and so they rot, causing huge "dead zones." Similar phenomena—which only jellyfish seem to enjoy—are occurring throughout the world, from the Baltic Sea to the Chesapeake Bay, and from the Black Sea in southeastern Europe to the Bohai Sea in northeastern China.[137] Our oceans, having nourished us since the beginning of the human species some 150,000 years ago, are now turning against us.

That dynamic will only grow more antagonistic as the oceans become warmer and more acidic because of climate change. Fish are expected to suffer mightily from global warming,[138] making it essential that we preserve as great a number of fish and fish species as possible so that those that are able to adapt are around to evolve and propagate the next generations of marine life. In fact, new evidence tentatively suggests that large quantities of fish biomass could actually help attenuate ocean acidification.[139] In other words, fish could help save us from the worst consequences of our own folly—yet we are killing them off. The jellyfish-ridden waters we are seeing now may be only the first scene in a watery horror show.

THE ROLES OF GOVERNMENTS

TO HALT THIS slide toward a marine dystopia, government intervention is required. Regulatory agencies must impose quotas on the amount of fish caught in any given year, and the way they structure such quotas is very important. For example, simply permitting all fisheries to catch a given aggregate number of fish annually results in a wasteful build-up of fleets and vessels as fisheries race to grab as large a share of the quota as possible before their competitors do. Such a system may protect the fish, but it is economically disastrous: the entire annual quota is usually landed in a short period, leading to temporary oversupply, which, in turn, leads to low prices. The alternative is to limit the number of fishers, with those retaining "access privileges" being able to catch their assigned fraction of the overall quota whenever they want, without competing against other fishers. Such individual quotas lead to less overall fishing effort and, hence, larger profit in the fishery.

Unfortunately, most fisheries economists, fixated solely on corporate short-term profits, argue that for such a system to work, access privileges must be handed out for free, be held in perpetuity, and be transferable (i.e., sellable and buyable like any other commodity). They call this construct "fishing rights," or "individual transferable quotas." However, there is no reason why a government should not auction off quotas with access privileges.[140] The highest bidder would secure the right to a certain percentage of the quota, and society as a whole would benefit from providing private access to a public resource. This would be similar to ranchers paying for the privilege to graze their cattle on federal lands. Grazing "rights," on the other hand, would simply give ownership of public land to ranchers, which is something few would consider.

Some Pollyannas believe that aquaculture, or fish farming, can ensure the health of wild fish populations without government action—a notion supposedly buttressed by statistics of the FAO showing such rapid growth in aquaculture that more than 40 percent of all "seafood" consumed now comes from farms. The problem with this argument is that China reports over 60 percent of the world's aquaculture production, and the FAO, which has been burned by inflated Chinese statistics before, expresses doubt about its stated production and growth rates.

Outside of China—where most farmed fish, such as carp, are freshwater vegetarians—aquaculture produces predominantly carnivorous marine fish, like salmon, which are fed not only vegetal ingredients but also fishmeal and fish oil, which are obtained by grinding up perfectly edible herring, mackerel, and sardines caught by what is coyly called "reduction fisheries."[141] Carnivore farming, which requires three to four pounds of smaller fish to produce one pound of a larger one, thus robs Peter to pay Paul. Aquaculture in the

West produces a luxury product in global terms. To expect aquaculture to ensure that fish remain available—or, at least, to expect carnivore farming to solve the problem posed by diminishing catches from fisheries—would be akin to expecting that Enzo Ferrari's cars rather than an emphasis on public transport can solve the gridlock in Los Angeles.

Others believe that fish populations can be rebuilt through consumer awareness campaigns that encourage buyers to make prudent choices. One such approach is to label seafood from fisheries deemed sustainable. In Europe, for example, consumers can look for the logo of the Marine Stewardship Council (MSC), a non-profit started by the World Wildlife Fund and Unilever, which has a large fish-trading division. At first, the MSC certified only small-scale fisheries, but lately, it has given its seal of approval to large, controversial companies. It has even begun to measure its success by the percentage of the world catch that it certifies. Encouraged by a Walton Foundation grant and Wal-Mart's goal of selling only certified fish, the MSC is actually considering certifying reduction fisheries, with the consequence that Wal-Mart, for example, will be able to sell farmed salmon shining with the ersatz glow of sustainability. (Given the devastating pollution, diseases, and parasite infestations that have plagued salmon farms in Chile, Canada, and other countries, this "Wal-Mart strategy" will, in the long term, make the MSC complicit in a giant scam.[142])

The other market-based initiative, prevalent in the United States, distributes wallet-sized cards designed to steer consumers toward fish that the group issuing the cards deems to have been caught sustainably. Their success is considerable if measured by the millions of cards given away, for example, by the Monterey Bay Aquarium, but assessing the impact on the fisheries is difficult. For one thing, the multitude of such cards leads to contradictions and confusion, as the same fish are assessed differently by different organizations. For example, ahi tuna (yellowfin) was once rated as "safe," "questionable," and "avoid" on the wallet cards issued by three different U.S. NGOs.

A bigger issue, however, is that these cards generate only "horizontal" pressure—that is, a group of restaurant-goers might chide each other for ordering the cod filet or might ask the overworked student who served them where the fish came from, but this pressure does not reach wholesalers, fleet operators, or supermarket chains. "Vertical" pressure exerted by environmental NGOs on such decision-makers is far more effective. But, if that is true, why not directly pressure the government and legislators, since they are the ones who regulate the fisheries?[143]

The fact is that governments are the only entities that can prevent the end of fish. For one thing, once freed from their allegiance to the fishing-industrial complex, they are the ones with the research infrastructure capable of prudently managing fisheries. For another, it is they who provide the billions of dollars in annual subsidies that allow the fisheries to persist despite the lousy economics of the industry. Reducing these subsidies would allow fish populations to rebuild, and nearly all fisheries scientists agree that the billions of dollars in harmful, capacity-enhancing subsidies must be phased out. Finally, only governments can zone the marine environment, identifying certain areas where fishing will be tolerated and others where it will not. In fact, all maritime countries will have to regulate their Exclusive Economic

Zones (EEZs, the 200-nautical-mile boundary areas established by the UN Convention on the Law of the Sea within which a country has the sole right to fish[144]). The United States has the largest EEZ in the world, and it has taken important first steps in protecting its resources, notably by creating a huge marine reserve in the northwest Hawaiian Islands. Creating, or re-creating, unfished areas within which fish populations can regenerate is the only opportunity we have to repair the damage done to them.

There is no need for an end to fish, or to fishing, for that matter. But there is an urgent need for governments to free themselves from the fishing-industrial complex and its Ponzi scheme and to stop subsidizing the fishing-industrial complex and awarding it fishing "rights" when it should in fact pay for the privilege to fish. If governments can do this, then we will have fish forever.

MAJOR TRENDS IN SMALL-SCALE FISHERIES[145]

SMALL-SCALE FISHING, GOOGLE, AND THE SCIENCES

ONE OF THE major trends in global fisheries is increased competition between small-scale and large-scale fisheries due to overfishing[146] and overcapitalization.[147] The two sub-sectors share numerous attributes across countries, although the scale of boats used differs. In developing countries, small-scale fisheries have smaller boats than in developed countries, where small-scale fisheries boats can be as large as industrial boats in developing countries.

One major result of the increased competition between the two sub-sectors, especially in developing countries, is the marginalization of small-scale fisheries. Although they meet most of the criteria required for an enlightened fisheries policy in employment and income distribution, fuel use, product quality and distribution, and sustainability,[148,149,150,151] small-scale fisheries are undermined by two related trends. One is the entry of landless farmers or cattle-less pastoralists into coastal small-scale fisheries, where they overwhelm the traditional fishers and local governance arrangements.[152] The other is the limited productivity of tropical coastal ecosystems,[153] which cannot support ever-increasing numbers of both traditional fishers and new entrants.

Ensuring the sustainability of coastal fisheries, especially in the tropics, requires that the migratory flows into coastal fisheries be reversed. This is a political problem, however, and it ought to be informed mainly by social scientists, not biologists. Yet fisheries biologists—I dare not speak of fisheries *scientists* in this context—share with fisheries economists the dubious privilege of being responsible for most of today's ideas about fisheries management.

Other social scientists, notably anthropologists, have far less input. This can be illustrated quantitatively: in 2005, if one entered "fisheries+ecology" into Google Scholar, one got, at the top, a paper with over 100 citations, while the subsequent papers also had noticeable impacts, as measured by citations (76,500 hits in total). This was similar for "fisheries+economics" (22,100 hits), as one would expect. But with "fisheries+anthropology" or "fisheries+sociology" as search terms, the top-ranking items were minimally cited and subsequent items remained uncited (3,260 and 3,950 hits, respectively).[154] Why is that so and what does this order-of-magnitude difference imply for fisheries research and policy?

This is an issue about which I cannot pretend to be neutral. I am a fisheries biologist, and I answer questions about the role of various fisheries-related disciplines from that perspective.[155] Moreover, I will not pursue inferences on the past of fisheries (before 1950), because these are the provinces of historical geography, archaeology, and ultimately paleontology, which will not be called upon here.[156]

THE POST-WWII HISTORICAL BACKGROUND

BY 1950, THE countries of western and eastern Europe and other developed countries had recovered enough from World War II that they could relaunch their strongly industrialized fisheries. At that time, the majority of what are now "developing countries" were still under European domination, or in the process of emancipating themselves from it, a process that mostly ended in the 1960s. As subsequent developments made clear, these countries were not only "underdeveloped" but also actively held back, as manifested in neocolonialist policies favoring continued economic links with the former colonial master.[157,158]

European and North American fisheries in the North Atlantic,[159] and those of North Asia in the Pacific, peaked in the 1970s, with all major fish populations exploited to the fullest.[160,161] Fishing fleets began to spill over from this region to areas farther south, such as West Africa, where they became the first distant-water fleets in those waters.[162,163] Simultaneously, there were, in numerous countries, a multitude of fisheries development projects, then advertised as a noble effort to help the newly independent countries of what became known as the Third World to make the best use of their marine resources. Many of these projects are now understood to have been the result of an East–West[164] rivalry that used bilateral aid projects as part of a containment policy, hastily abandoned—along with the projects—after the collapse of the Soviet Union.

These development projects, usually staffed by biologists, largely neglected the experience of traditional fishers in the countries concerned, as has been documented in various classic papers,[165] in compilations of case studies,[166,167] and in historical accounts.[168] Rather, these projects sought to create local industrial fisheries based mainly on trawling and in direct competition with more traditional forms of fishing.[169] These were success stories in the sense that huge fisheries were developed, notably in Thailand,[170] which itself initiated distant-water fisheries.[171] In Africa, however, such projects failed to induce the development of large-scale fisheries, despite the presence of favorable starting conditions in some countries, notably Ghana.[172,173]

The primary reason for this was the existence of distant-water fleets. Along much of Africa's coastline, especially off West Africa, there is still direct competition between local small-scale fishers and foreign industrial fisheries (that is, distant-water fleets). In most countries of Latin America and the Caribbean, the industrial sector often consists of national fleets exploiting large pelagic fishes and hence not directly competing with near-shore, small-scale fisheries. In some other countries, distant-water fleets generate conditions similar to those in West Africa.[174] Whatever the route that "development" took, the goals of fisheries development were generally "biological" (high catches, utilization of all resources, etc.), to the nearly complete neglect of social goals such as employment, community well-being, and food security, among others.[175]

These events and trends should have provided many opportunities for social scientists to contribute to the discourse in fisheries science, and even to insert themselves in actual fisheries management and policy making. This generally did not occur, as shown by the numbers from Google Scholar given in the previous section. I think this failure is due to two major aspects of the "research mode" of social scientists. To put things stereotypically, social scientists working on fisheries (1) neglect key quantitative variables (this is especially true for the catch of small-scale fisheries, which social scientists could often estimate reliably, given their access), and (2) fail to propose and test models of social behavior of sufficient generality to be useful for policy making.

I will not back these claims through exhaustive citations of the literature, because proving negative statements of this sort would involve detailed hermeneutics of the key texts in social sciences and fisheries. Rather, I will give examples of the two items above and comment on them.

MARGINALIZATION I: CATCH UNDERESTIMATION

THE FOOD AND Agriculture Organization of the United Nations (FAO) issues world fisheries statistics annually. Many researchers take these statistics to refer to the world catch, but these statistics are incomplete. Discarded bycatch is not officially reported, although the FAO itself has commissioned estimates of discards,[176,177] which others have commented on.[178] Also, a significant portion of the catch is landed illegally and not estimated.[179] Finally, many fisheries are unregulated, either because they take place on the high seas or because they are small scale and fall below the radar of the national statistical agencies that report national catches to the FAO.[180,181]

This issue may be illustrated by the fisheries of the South Pacific, which tend to consist of two sub-sectors: the tuna fisheries,[182,183] which mainly involve distant-water fleets operating in the Exclusive Economic Zones (EEZs) of the various countries, and what might be called the inshore fisheries, based on exploiting reef and other neritic fishes.[184] As indicated by the FAO statistics and reflected in the database of the Sea Around Us, which presents these statistics in a geographic context (see http://www.seaaroundus.org), numerous South Pacific countries report no, or very limited, inshore catch, although they are known to depend heavily on locally caught fish for their nutritional needs and food security.[185,186]

Social scientists are well placed to contribute estimates of these small-scale fisheries catches, because of their local contacts and because they are often embedded in the very institutions that take the pulse of local small-scale fisheries. And social scientists should know the importance of catch levels, which are what make people go fishing.

Yet, *Words of the Lagoon*,[187] a classic whose vivid description of Palau's reef fishery many fisheries anthropologists rightly attempt to emulate, does not contain the catch (and catch per fisher) data that, much better than words, would allow evaluation of the extent to which traditional Palauan fishing practices are sustainable and an assessment of the role of the reef fishery in the rural economy of Palau.

How important such a role can be, and the extent to which it is underestimated by FAO statistics, may be illustrated by American Samoa, whose reconstructed inshore fisheries catch for the period 1981 to 2002, although based on conservative assumptions,[188] was seven times as high as the official statistics reported to FAO. This higher catch also contributed nine times more to the rural economy than originally assessed.[189]

These results, which can be reproduced throughout the region,[190] imply that the emphasis currently given to tuna in the South Pacific region may be misplaced, along with a version of food security that uses access fees paid by distant-water fleets to import fish and other food (notably Spam[191]). This emphasis has, at its flip side, the nearly complete official neglect of the inshore small-scale fisheries and their marginalization.

MARGINALIZATION II: MALTHUSIAN OVERFISHING

THE NEXT ISSUE concerns models, mental constructs meant to reflect important aspects of reality, such as our knowledge of it, and to enable the exploration of, for example, the implications of certain policies affecting that reality.

I assert that social scientists have rarely proposed generalizable models of fishing communities. Rather, they have tended to offer descriptions of localized situations, from which such models might be constructed, and against which they could be tested (since all the nontrivial assertions of such models should be treated as hypotheses).

Construction, articulation, or eventual refutation of such models, the most successful of which become "theories," is currently practiced in the natural sciences. The following, although not expressed in quantitative terms, and dealing with social science issues, is structurally akin to such models. This model describes what I called, perhaps unfortunately, the Malthusian overfishing of small-scale fisheries, although raw population growth is only one of its drivers. The major elements of this model, each formulated as a testable hypothesis, are as follows:

1. A large agricultural sector (at least when compared with the fisheries sector) releases excess labor because of population growth, mechanization, and land "reform."
2. These landless farmers migrate to urban, upland, or coastal areas.
3. Under this influx, traditional arrangements preventing open access to the fisheries gradually collapse.
4. This collapse leads to excessive fishing pressure.
5. That pressure is exacerbated by inshore industrial fishing.
6. It is also exacerbated by new recruits to fishing, as the male children of fishers pick up their fathers' trade.
7. As many young women leave their communities to work in urban areas, and thus provide a subsidy for men to continue to fish even when resources are depleted, fishing pressure increases even more.
8. The migrants to upland areas accelerate or complete the deforestation initiated by logging companies, which leads to siltation of rivers and streams.
9. Eventually, this siltation smothers coral reefs and other coastal habitats, thus further reducing the yields of coastal fisheries.[192]

Since this model was formulated, my personal observations in, and literature from, South America, South and Southeast Asia, and Africa suggest that these trends have been accentuated. Therefore, I would suggest that this model might still provide good questions for fisheries anthropologists and other social scientists to investigate and test. However, one new element is the globalized export markets into which communities of small-scale fishers can plug themselves directly and which offer them potentially higher incomes. But these markets also contribute to the removal of the last remnants of traditional, place-based management.

MALTHUSIAN OVERFISHING: THE 2004 TSUNAMI

FOLLOWING THE TSUNAMI of December 26, 2004, which devastated South and Southeast Asia, the Malthusian overfishing model provided the background for recommendations on how to mitigate the damage,[193] as follows:

> The tsunami that hit South and Southeast Asia on December 26, 2004, taking a horrific toll in human lives, also affected several coastal industries, including tourism and agriculture, though to what extent is unclear. In some areas, these effects were exacerbated by existing environmental problems stemming from settlement and industry.[194] The governments of Thailand and Indonesia have announced some estimates of fishing boats lost and highlighted the need for investments to restart the fisheries.
>
> However good their intentions,... Western aid agencies, and indeed, the governments of the region would be ill-advised to rebuild the fisheries as they were before the tsunami. Apart from oceanic fisheries for tuna and other large fish, fisheries in the tsunami-affected region

fall into two categories: "artisanal fisheries," relying on small (5 meters or less), owner- or family-operated craft, some non-motorized; and "industrial fisheries," using larger vessels, mainly trawlers but also other specialized craft with salaried crews. Jointly, their fishing activities have radically depleted the nearshore resources, down to depths of 100 meters in places. Governments in the region have tried to encourage the industrial fisheries to operate farther offshore, but with little success, mainly because biological production in tropical waters is much higher inshore than offshore.[195] Hence, the artisanal and industrial fisheries essentially target the same shrimp and fish, leading to intense competition.

This competition and the ensuing violence, including boat burnings and riots, can be serious enough to prompt governments to take action, such as the 1980 ban on bottom trawlers in western Indonesia.[196] Usually, however, government policies ignore these conflicts. Sometimes they exacerbate conflict by subsidizing the construction and operation of industrial vessels, even in cases where these do not add to the total catch, but reduce that of the artisanal fishers. International aid has often aggravated this through technological and capital transfers, or donations of surplus vessels. Meanwhile, failed agricultural and social policies aggravate the situation by driving thousands of landless farmers to coastlines, where they usually fail to emulate the more sustainable ways of "traditional" fishers.[197]

After the tsunami, the initial push will be to get people back to the jobs they know, and it will be hard to argue otherwise in the midst of the chaos. But rebuilding the fisheries without structural reform will only intensify these trends and conflicts. The challenge is to rebuild fisheries while directing as much money and energy as possible to generating land-based job opportunities for young fishers. Emphasis should be given to basic education and technical skills: many fishers in South and Southeast Asia are illiterate, and this limits their social mobility.[198] Amending the old adage that teaching people to fish is better than giving them a fish to eat, we should instead be teaching them to repair bikes, sewing machines and water pumps.

This advice was not heeded:[199,200,201] the urge to subsidize was just too hard for relief agencies to resist, never mind the devastating effects of subsidies, which can destroy small-scale fisheries as well as industrial ones.[202,203]

MALTHUSIAN OVERFISHING AND RESEARCH

GOOGLE SCHOLAR SUGGESTS that the first paper in which the Malthusian overfishing model was fully developed,[204] following briefer accounts in various outlets, was (as of January 2006) cited over 30 times, but overwhelmingly by biologists.[205] The major criticism that the Malthusian overfishing model has received from social scientists is that industrial fisheries are considered the cause of the phenomenon because they massively deplete the resources previously available to small-scale fishers. This is partly true: most small-scale fisheries throughout the world have seen their resources depleted by industrial fishing vessels, notably trawlers, fishing on or near their inshore fishing grounds. Yet this is not the whole story. An ever-increasing number of small-scale fishers operating motorized canoes or similar mobile and versatile crafts can deplete the entire resource on a country's continental shelf, and this is particularly clear in West Africa, the Caribbean, and the South Pacific. The local and foreign industrial fleets are only accelerating—albeit to a tremendous extent—the overfishing, which, if not controlled, would eventually engulf even the most seemingly benign small-scale fisheries.

Thus, for example, the foreign fleet exploiting the Bijagós Archipelago in Guinea-Bissau consists of not only the ubiquitous trawlers of European fleets but also fishers on motorized canoes from neighboring Guinea and from as far away as Senegal. The latter set up camps in the Bijagós' outlying islands, from which they systematically deplete all resources in the neighborhood before returning with their catches, to be landed in Conakry or Dakar.[206]

The best example of Malthusian overfishing is, however, the Bolinao reef fishery, in the Philippines, which has been documented in great detail.[207] In this instance, internal dynamics led to the destruction of the resource without any help from a large-scale fishery. Perhaps I should mention that it is also the site that inspired the model.[208]

These observations can be interpreted in numerous ways, and, as stated in the introduction, I can offer only the perspective of a fisheries biologist. With these caveats, here are my views on research topics that I consider worthwhile.

FOR FISHERIES ECONOMISTS

FIRST, RIGHTS-BASED FISHING seems to be *en vogue*. However, in most of the developing world, rights-based fisheries likely will not work, at least not in the form of individual transferable quotas (see the essay titled "ITQs: The Assumptions Behind a Meme"). In fact, restricting entry for small-scale fishers (and even for national industrial vessels) is not going to be politically feasible until distant-water fleets are curtailed or at least become invisible—in other words, until they are far offshore.

Second, the foreign exchange obtained from distant-water fleets, while music to the ears of most developing-country politicians, does not necessarily lead to economic development. A complete analysis ought to always look at whether or not this foreign exchange makes its way into the local economy. If not, the economic role of local fisheries, which generate substantial rural incomes, will be more important.[209]

FOR ANTHROPOLOGISTS

REPORTING ON THE culture of a village, including its local adaptations to the material environment and emphasizing their uniqueness vis-à-vis other villages and the mainstream culture of a country is how one gets credibility in a field where the "local" is so prized.[210] Yet the negligible role that anthropology and related social sciences play in informing fisheries management should be a warning that the social sciences must generalize from local conditions to formulate people-oriented and sustainable government policies for an entire region or an entire country, not just local communities.

The reason biologists and economists have come to almost monopolize policy is because they are willing to develop such generalizations even when they don't understand the social consequences or the assumptions about human behaviors on which these policies have been developed.

Furthermore, there is a real danger that if the small-scale fisheries don't manage the influx of new fishers into their midst, they will be destroyed by new entrants—in this case, nontraditional fishers. Thus, there is a real need for generalizable work on the causes of migration within, into, and from coastal fisheries.

FOR SOCIOLOGISTS

THERE ARE VERY few studies available of how collapsed fisheries are phased out and how this affects different social groups. Yet, if the trends alluded to above hold, "fisheries extinctions" should become common. We need guidance on how to integrate ex-fishers into other gainful activities and how to create other sources of employment in communities that cannot continue to rely on fishing. To date, no such guidance is available, though there is a great need for it, as shown by the 2004 tsunami.

Another topic that is under-studied is the role of women in small-scale fisheries. Women catch fish, though often not the glamorous ones,[211] and they process fish, too; these activities are conventionally studied. What is little studied (if at all) is how women (the wives, sisters, and daughters of fishers) who work outside the fishing sector and keep the family in cash subsidize male fishers and allow them to continue exploiting overfished resources.

FOR ALL OF US

THE MASSIVE REDUCTION of biomass, which is the characteristic *modus operandi* of modern fisheries,[212] and the erosion of biodiversity and ecosystem function that this entails, as expressed by the fishing down phenomenon,[213] endanger the long-term sustainability of fisheries. In the long term (two to three decades?), fisheries and fishing-based cultures will not survive if we do not manage to put small-scale fisheries and resources first and to rein in both the floating behemoths that industrialization has brought us and the massive rural migration into small-scale fisheries. Realistic scenarios for such transitions exist,[214] but the alternative scenarios, with more overfishing by subsidized industrial fleets and neglect of the small-scale fisheries, are still more appealing to policy makers.

I conclude with a vision of vibrant small-scale fisheries contributing to coastal communities and supplying, throughout the world, the bulk of fish for human consumption, harvested with a minimum expenditure of energy, in a sustainable fashion.[215] This is what small-scale fisheries can do, once they are freed from the constraints under which they now operate.[216]

ITQS: THE ASSUMPTIONS BEHIND A MEME[217]

INDIVIDUAL TRANSFERABLE QUOTAS IN THEORY

THIS CONTRIBUTION COMMENTS on four papers on individual transferable quotas (ITQs) published in a 1996 issue of *Reviews in Fish Biology and Fisheries*, which jointly presented a comprehensive review of one approach to fisheries management. That approach could be described by Thomas Kuhn's concept of a "paradigm shift," had the term not been so overused as to be drained of meaning. This approach is included here because ITQs, now known as "catch shares," are very much *en vogue*.

The spread of ITQs, bursting as they did from the periphery to the center stage of fisheries management, and from small countries such as Iceland and New Zealand, suggests an evolutionary metaphor, even though we are not dealing here with the emergence and spread of a new gene but with one of Richard Dawkins' "memes," an idea that competes with other ideas for our attention and space in our brains. I shall return to this metaphor following a discussion of some implicitly held assumptions of critics and proponents of ITQs, assumptions that I believe should be made explicit. R. Quentin Grafton[218] tells us, in a concise review of the theory and practice of ITQs, that this idea originated in the 1960s with the "pollution quotas" now widely used by power utilities to rationally allocate the resources they have for reducing the amounts of pollutants they emit.

Crediting Francis Christy with being one of the first to look through the clouds of indignation these quotas generated and to see their potential for fisheries management, Grafton then describes how ITQs, by making fishers the owners of the harvestable part of a fish resource, can help overcome the perennial problems of fisheries: the "race to fish," the resulting overcapitalization, and the assault on the resource that usually ensues. The case for ITQs—and Grafton's case—seems clear-cut and would be convincing to anyone were it not for various, generally unstated assumptions, which, at least in my limited experience, tend to poison discussions about ITQs.

The most common assumption about ITQs is that they are part of some right-wing plot to privatize the sea by handing over its fish resources to big corporations. This view is commonly held not only because we live in an age when belief in conspiracies is rampant, but also because big corporations already own big chunks of the Earth's natural resources. To imagine that they should be able to skew to their advantage—and to the detriment of, say, deckhands or even small owner-operators—the process by which ITQs are initially allocated does not require much fantasy.[219] Not explicitly addressing this issue is glib and is likely to justify the fear behind it.

A related concern is concentration of quotas in a few hands following their initial allocation. Grafton addresses the concentration issue by simply noting that "whether such an adjustment takes place [...] is entirely dependent on the characteristics of the fishery."[220]

INDIVIDUAL TRANSFERABLE QUOTAS IN PRACTICE

IN HIS DESCRIPTION of the evolution of Iceland's ITQ system, Ragnar Arnason[221] also mentions factors that should prevent concentration of ITQs in that country. Some of these are implicit, as a result of the structure of the Icelandic economy, and some are explicit, in the form of legislative acts.[222]

In any democratic society, it is ultimately the electorate that must decide to whom common-property resources should be handed over, and with what restrictions. Thus advocates of ITQs should stress the right of the electorate to effectuate such transfers of property, through elected governments, rather than limit their case to restating their economic-efficiency mantra.

Another assumption that is frequently made by fisheries economists is that communities of small-scale fishers, or of other marginalized groups with claims to fisheries resources, cannot be owners of ITQs, as seemingly implied by the "I" of ITQ. But they can, and this is well illustrated by the case of New Zealand, where the Maori community, which otherwise does not own much of what was once their country, now controls 40 percent of the ITQs.

The ability of communities to become significant owners of fisheries through ITQs will obviously depend on their share in the initial allocation process and on their ability not only to hold onto their initial ITQ (which may be mandated) but also to concentrate their resources—meager as they might be—on the acquisition of further ITQs. One can even imagine a situation wherein such communities—for example, because of a strong fishing tradition—might decide to acquire ITQs at prices that corporations without such a tradition, and with only short-term returns in mind, might be unwilling to consider. It is unfortunate that so little of the attention of ITQ advocates appears to be devoted to issues of this sort, although they might well determine the acceptability of ITQs in areas where the voice—electoral or otherwise—of small-scale fishers and/or of marginalized groups is strong enough to tip the balance of the decision.

Another series of assumptions concerning the research needed to implement and maintain an ITQ system stems from the emergence of ITQs at a time when governments are not only supposed to privatize the natural resources they control but also expected to "downsize themselves"—to reduce their role in managing resource exploitation by the private sector. New Zealand is the country that most readily comes to mind.

The irony is that output-based regimes (such as those relying on ITQs) turn out to be far more management-intensive than the input-based regimes (e.g., those controlling fishing effort) they are meant to displace and that government scientists can be expected to perform most of the "stock assessments" needed to generate the total allowable catches (TACs) upon which the whole ITQ system rests. John Annala,[223] among others, demonstrates the relatively high cost of the scientific research required by an ITQ system, although various tricks are used in New Zealand to make stock assessment scientists look like they are not working for the government.

The requirement for "much more accurate and timely stock assessment" is the aspect of ITQs that Carl J. Walters and Peter H. Pearse[224] emphasize, because in the absence of reliable data, "quotas may need to be so conservative that foregone catches could wipe out the economic gains from quota management." They then present a highly original approach to defining optimum fishing mortality—for example, by incorporating uncertainty about the size and reproductive capacity of exploited fish populations and showing *en passant* that the TAC for the northern cod of eastern Canada was far too high, making its collapse inevitable.

Turning around the argument of Walters and Pearse, one can thus conclude that an ITQ-based regime will not protect fish populations better than input-based regimes if governments (or quota owners) do not invest in more and better fisheries science.

Further, even good science is not going to help in the absence of comprehensive monitoring schemes (including on-board observer programs) to suppress quota busting, high-grading, and other nefarious practices.[225] Annala's description of this, and of the specialized government accountants who, in New Zealand, audit the accounts of ITQ holders, should convince anyone that assuming ITQs will reduce the role of governments is not realistic.

The preceding discussion of some assumptions pertaining to ITQs comes from my close reading of the four key papers in question. A looser reading of these same papers reveals the evolution of the ITQ meme.

It was rightly noted[226] that "in practice, the United States is a tough place to tell somebody he or she can't go fishing while the rest of the fleet is still grinding away on the grounds." It has been suggested that to institute access-based schemes such as ITQs, "you need a fishery where nobody has recourse to the U.S. courts to challenge the new system." Similarly, in the European Union, every potentially effective plan for reducing fishing effort or limiting access is countered by nationalistic lobbying coupled with threats that the country will leave the Union or by demonstrations that often turn violent, or both. Thus the meme for ITQ did not initially spread in a major country such as the United States, although that is where it originated.

Rather, for a meme such as ITQ to establish itself as dominant, a relatively small, isolated community is initially required, in which a rational debate about issues could occur, and in which the risk of failure is understood by most, a risk, moreover, that has to be borne by the entire community. Amason[227] puts it simply: "the Icelanders simply cannot afford to run an inefficient fishing industry." (The United States can, and so can the European Union.)

Evolutionary biologists will appreciate the similarity—in this context, between Iceland and a small, isolated biological population—of a meme establishing itself in a marginal site that could not expand into a larger population (for example, the United States or the European Union). In fact, the similarity becomes stronger when one realizes that Iceland became able to consider the utility of ITQs only after it had, rather militantly, expelled all foreign fishers from its 200-nautical-mile Exclusive Economic Zone. (Then another new meme, which, as an instant of reflection will suggest, probably co-evolved with the ITQ meme.)

From its strong base in Iceland, New Zealand, and gradually Australia as well, the meme for ITQs now stands poised to spread into the core population of fisheries managers in temperate zones in Europe and North America. Yet, as would be true for any gene, its victory is never assured: Arnason[228] tells us, for example, that in Iceland, partisans of the old, input-control meme still hope for a reversal of their fortune.

I conclude with what may be an expected pitch for the tropical and subtropical multispecies fisheries, which contribute a large fraction of the world's catch. In spite of Francis Christy's early forays in this area,[229] I have never seen or heard an argument, by an ITQ advocate, on how to apply this tool to these species-rich but data-poor fisheries. What may be the implicit assumption behind this—that these fisheries are somehow irrelevant? Fortunately, the answer—whatever it might be—does not matter, because, in fact, ITQ-based management systems should have a hard time with such fisheries and in countries the developing status of which is largely defined by the absence of the strong administrative and research infrastructure required for output-control management.

Thus, marrying my two themes—the implicit assumptions of ITQs and their metaphoric interpretation as meme—I predict that the ITQ meme does not tolerate warm water.

PUTTING FISHERIES MANAGEMENT IN ITS PLACES[230]

TOWARD A PRIVATIZATION OF THE SEA?

EXCEPT, MIRACULOUSLY, FOR Jules Verne's, scientific predictions always turn out to be wrong. However, the third millennium has come, fisheries resources are going, and it is impossible to resist the urge to make a few predictions about the future of fisheries management and of fisheries research as a scientific discipline. And no, the s at the end of the title of this essay is not out of place: In the future, fisheries management and its associated science will have to deal with "places" far more than they have in the recent past. In many cases, they will have to return to ancient modes of allocating fisheries resources to local communities, rooted in physical places.

As discussed in the previous essay, the trend now is toward privatization of fisheries resources through individual transferable quotas (ITQs) and similar instruments, and there are also attempts to privatize the research scientists and the detailed assessment work that these instruments require.[231,232] Hopefully, this trend will reverse itself when more people realize that self-interested exploitation schemes, while eminently compatible with the acquisitive mood of our times, do not resolve the discrepancy between human and natural time scales any more than do the open-access schemes they might replace.[233]

Many exploited fish species—for example, fish in temperate waters and large predators on coral reefs—are long-lived, with natural mortalities of 10 percent per year and often less.[234] This fact implies that for exploitation to be sustainable, fishing must not extract more than about 10 percent of a population's biomass per year, especially where data are sparse.[235] Even this low percentage is enough, however, to quickly remove accumulations of large, old females—the source of most of the eggs and subsequent recruitment of long-lived fishes. The relationship between fish size and egg production is highly nonlinear, with large females being far more fecund than an equivalent weight of small ones. This nonlinearity is so pronounced that, for example, one single ripe female red snapper, *Lutjanus campecheanus*, that is 24 inches long and weighs 27.5 pounds contains the same number of eggs (9.3 million) as 212 females that are each 16.5 inches long and weigh 2.4 pounds each.[236]

The massive reduction of egg production that occurs when large females are removed is one of the reasons why exploited fish populations fluctuate as much as they do, regardless of the direct effect of environmental fluctuations.[237,238] F.I. Baranov, one of the founders of fisheries science, was perhaps the first to realize that "by reducing the fish population, fishing itself [generates] the [population growth] increment which, in turn, sustains the fishery."[239] Let's not wait too long to admit that, similarly, fisheries also generate many of the fluctuations that beset fish populations, including the occasional collapse—the ultimate fluctuation.

Further, even a low fishing pressure, when caused by gear such as a bottom trawl, will have profound effects on the habitats of demersal fish species, notably by eroding often century-old bottom structures such as "oyster reefs," sponge communities (e.g., of *Poterion* in South East Asia), and other beds of sessile, filtering organisms. The result is increased water turbidity and a gradual transition, within coastal ecosystems, from a demersal to a pelagic food web—a very common type of transition.[240]

MARINE RESERVES AS PLACES OF RECOVERY

THUS, MY CONTENTION is that even very low rates of fishing mortality cannot be sustained by demersal fish populations unless a sizeable fraction of their spawning adults are completely inaccessible, owing to some natural refuge (underwater canyons, large boulders, etc.). These refuges are the very spots that good fishers must discover and drain if they are to maintain high individual catches—whatever the average level of fishing pressure. But can we reconcile the vastly different time scales of humans on the one hand, and those of fish and of benthic communities on the other? Not through application of "optimal" rates of fishing over large areas, however detailed the studies that led to these rates. Rather, fish (including major commercial species) require new refuges—marine reserves—providing shelter for a wide variety of species, which will then be protected from collapse.

For this to work, though, there must be agreement not to fish in certain places. That can happen only if those who abstain from fishing within marine reserves accept the rationale for those reserves and benefit from their existence—that is, if a sense of place re-emerges within fishing communities as they become the local guardians of fish populations and not their roving executioners. Such agreement may emerge if science continues to confirm that suitably placed and suitably sized marine reserves will perform what is expected of them.[241] And, perhaps not surprisingly, given the wide scope of Beverton and Holt's classic work, when assessing the suitability of places for marine reserves or their sizes, we will be using ideas implicit in that work. We will also use their work in explicitly dealing with marine reserves, because yes, they also dealt with those.[242,243]

Although sometimes tempted by pessimism, I hope that we humans will find ways to match our numbers and our demands with what our planet can provide (which is not the case now). This will require that we abandon rape and pillage as our major mode of interaction with natural resources. For fisheries, it will require rediscovering places for fisheries management.

FISHERIES MANAGEMENT: FOR WHOM?[244]

WE ALL PASS THE BUCK

THE CRISIS OF fisheries is real, and it is global. It is an ecological crisis because fisheries production systems worldwide are losing their productive capacity. It is a socioeconomic crisis because industrial fisheries now rely on billions of dollars in government subsidies each year while simultaneously undermining the livelihood of millions of small-scale fishers.[245]

It is also an intellectual crisis because fisheries science has lost much of its hard-won credibility, partly because it still considers narrow industry interests the only legitimate "clients" for its services,[246] despite numerous fisheries-induced collapses, and many fisheries scientists are apparently unwilling to rely on the broad ecological knowledge so far accumulated to support management approaches based on precautionary principles, both in the tropics[247] and in cold temperate areas.[248]

It is finally an ethical crisis, and certainly one of alienation as well, as much of the fisheries sector has discarded, along with millions of tons of bycatch every year,[249] all notions of guardianship of the resources on which its survival depends.

In effect, the fisheries sector has abdicated this role to conservationist organizations, which industry representatives, from the CEOs of fishing corporations to lobbyists for sport fishing, too often criticize for their lack of "realism."

Canada's fisheries provide examples of all the abovementioned ills. Canada also offers one of the few instances of major fish populations being driven into collapse by a fishery that largely followed mainstream scientific advice.[250] In addition, Canadian fisheries reproduce numerous aspects of the global fisheries crisis, as they pit a corporate fishing fleet against owner-operated craft, both against sport fishers, all against the First Nations, all of these users against the Canadian Department of Fisheries and Oceans (DFO), and the entire sector against the taxpaying public at large—the ultimate sovereign. A detailed analysis of this buck-passing exercise on the Pacific coast of Canada exists,[251] and thus Canada can serve as a microcosm of the world's fisheries. In this essay, I thus alternate between Canadian and other examples when discussing solutions proposed to address these problems.

DEALING WITH REAL PEOPLE

THE MAIN THEME of this essay, written from the perspective of a fisheries biologist, is my contention that to be successful, future management schemes, whether based on market incentives, on co-management, or on governance arrangements, must involve local communities living in real places and exploiting fish populations that have places as well. However, before we look at the places of fish, we should discuss at least one of the half-truths that pollute such a discussion.

The cliché often used by fisheries scientists to counter conservationist arguments is that "management is not about fish but about people." This statement may resonate nicely in empty heads, but it implies that we fisheries scientists are or should be equipped to deal with "people issues." Our collective inability, nay unwillingness, to engage in collaborative work with social scientists (sociologists, anthropologists, historians) and our tendency to reduce what they commonly use as fact or evidence to the status of anecdote[252] show our discipline to be conceptually ill equipped to deal with people issues. Moreover, and perhaps more importantly, applied sciences dealing with real people (as opposed to our nominal fishers and hypothetical managers), such as medicine or psychology, have by necessity developed strong codes of conduct to regulate the ethical aspects of their interactions with people.[253] Thus, psychologists must obtain informed consent, usually in writing, before performing even seemingly harmless teaching experiments on graduate students, and complex protocols regulate the conduct of medical experiments on both animals and people. When human beings are involved, experiments on the efficacy of a potentially life-saving drug must be interrupted when the subjects receiving placebos can be shown to suffer from not being included in the treatment group. This and similar ethical issues, in many medical schools, are part of the curriculum and are taught by faculty specialized in medical ethics.

How about fisheries scientists? Do we really want to deal with real people and become personally responsible (as medical doctors are) for what we do, for the advice we give, and for the consequences of such advice?

TYPES OF SOLUTIONS

THREE CLASSES OF approaches have been proposed to address the global fisheries crisis and its local manifestations, usually as a complement or an alternative to national or international "top-down" regulatory approaches. These three classes are market-based approaches, community-based approaches, and ecology-based approaches.

MARKET-BASED APPROACHES

THE OPEN-ACCESS NATURE of most fishing grounds, in Canada and elsewhere, has long been identified as the major cause of the "race for fish" and its attendant ills, such as overcapitalization, overfishing, and population collapses. Consequently, numerous economists attribute these problems to "market failures," the inability of the market to properly account for, or "internalize," the social and environmental costs of fishing.[254] In contrast, there are far fewer accounts by economists of the active collusion (what other word is there?) between governments and large fishing enterprises, as can be seen in the boat-building and other subsidies that have enabled large fishing enterprises to exploit coastal resources far from their home ports[255] and, in the process, to marginalize otherwise efficient, localized, small-scale or artisanal fisheries.[256] Thus, it is not surprising that the market-based mechanism proposed to overcome market failures—individual transferable quotas (ITQs)—is often perceived not as a tool for resource management *per se* but as a ploy for transferring more public assets from public to corporate ownership.[257] When implemented, however, ITQs seem to achieve much of what is expected of them, reintroducing rationality—albeit in its narrowest, economic sense—into an industry that had gone irrational several decades earlier.[258]

COMMUNITY-BASED APPROACHES

TO EMPHASIZE COMMUNITY-BASED approaches here may appear to imply that they are something new—yet we know that in earlier times, and for obvious reasons, local communities were the only entities required for and capable of managing fisheries.[259,260] The reason community-based management is not a trivially obvious thing to do is that, in parallel with the development of large fishing boats (typically bottom trawlers, but also purse seiners and other types of industrial craft), national governments in recent decades have created centralized agencies to regulate fisheries, with an internal culture, and often an explicit mandate, that has favored industrial (and distant-water) fleets over smaller, locally based artisanal fleets.[261,262,263] The resulting inequities and alienation, in both developed and developing countries, are well documented in the literature and have spawned the concept of co-management, a sharing of responsibility between national governments (usually represented by a regulatory agency) and fisher communities.[264,265] Co-management, as currently conceived, may range from the right of communities to be consulted during a decision-making process to their nearly complete autonomy (with respect to fishing). Usually these arrangements assume that owners of fishing vessels have a legitimate interest in co-managing a fishery, but not their salaried crew or their wives, who are often involved in processing[266] the catch—an additional source of inequity and alienation.

Moreover, and perhaps even more importantly, these arrangements all imply that the resource users are the only group that governments need to consider—non-user groups are not usually seen as legitimate stakeholders in the management and allocation of resources. In contrast, the concept of "modern governance,"[267] wherein governments unable to accommodate all demands by particular user groups involve groups with different or even opposing interests in the decision-making process, thus forces the user groups—here, fishers—to justify their privileged access to public resources. In the case of coral-reef fisheries, this approach might imply, for example, the creation of local management councils in which fishers must negotiate with the operators of scuba dive resorts, and perhaps conservationist groups, and where government representatives only set and enforce rules for intergroup negotiations.

Governance arrangements of this sort, which would always be local, may result in levels of exploitation that are sustainable and compatible with the interests of different stakeholder groups, while also reducing transaction and enforcement costs for the central government.

ECOLOGY-BASED APPROACHES

THE IDEAS THAT Nature is "complex," particularly the oceans, and that we know very little about the processes producing and maintaining fishery catches are part of the smokescreen behind which disciplinary irresponsibility can be hidden. Actually we do know—and have known since the beginning of the 20th century, when F.I. Baranov developed the principles of quantitative fisheries science,[268] or at least since World

War II and the giant fisheries closure that it entailed[269]—that excessive fishing reduces fish populations and eventually causes them to collapse but that reducing fishing is sufficient, in most cases, for them to recover (given time).

What has enabled our historic fisheries to last for centuries is that a good part of each of the exploited populations was not susceptible to fishing, because it had access to natural refuges. The refuge in the case of northern cod was depth: in the past, the fishery occurred mainly inshore, and the old, large females whose reproductive output maintained the recruitment of young fish to the adult population were largely inaccessible to the inshore gear dominating the fishery.[270] Similarly, most tuna could be caught only if some of them strayed inshore, the rest of their population remaining safe in oceanic waters.

In the last decades, technological developments—powerful echolocating devices, extremely precise satellite positioning, and new gear—have made it possible to locate and catch almost any fish, anywhere, and to exploit what initially were refuges. It is beyond the scope of this essay to document this change. However, the well-documented recent collapses throughout the world would provide much of the evidence were it to be presented here.

Countering this creeping invasion of natural refuges is possible, and this possibility is probably what lies behind the growing scientific consensus about the efficacy of marine protected areas (MPAs), particularly in their most effective form, as "no-take" MPAs.[271,272] Essentially, no-take MPAs work as artificial refuges by reconciling Nature's time scales with those of fishers and markets as required, because even very low fish mortality can drastically reduce the numbers of the large, old, and highly fecund females that contribute most to recruitment. Thus, within a (suitably located and sized) no-take area or zone, the numbers, and eventually the biomass, of one (or several) previously decimated population(s) can recover, even as (regulated) fishing continues outside that zone. Gradually, as density and mean ages increase within the zone, it starts exporting eggs and eventually juveniles and adults that contribute to the adjacent fishery, soon offsetting through this export the fishing lost because of the no-take area itself. Tropical examples documenting the reality of these processes exist,[273] as well as examples from colder waters.[274,275]

The most important aspect of MPAs, however, may be that they simultaneously reduce risk and accommodate imperfect knowledge[276]—an intractable problem with quota-based or other traditional forms of management.[277] Contrary to much of what one may hear concerning our lack of knowledge about the dynamics of exploited populations, this cliché speaks not against but for the establishment of MPAs. Do engineers use lack of knowledge about the average weight of *Homo sapiens* as a reason for or against building elevators that can accommodate more than their stated maximum number of passengers?

OUR FUTURE IS IN PLACES

THE TEXT ABOVE should have made my sympathies clear: I believe that fisheries management, if it is to lead to anything sustainable, must take into account the places of people in its logic. It must consider, far more than has hitherto been the case, the places of small-scale fisher communities but also of other stakeholders and the places of fish, especially places where their populations can recover from fishing. Taking places into account will not, and indeed could not, be done by fisheries managers alone. Rather the public at large, which ultimately owns the resource and whose taxes have so far been misused to subsidize the carnage, must become involved. There are scattered signs that this is happening, throughout the world, as a result of the recent massive coverage of global fisheries issues in science- and Nature-oriented magazines,[278,279] in the general press, and as a result of the trend toward greater accountability of government agencies. The public will not become involved *en masse*, obviously. Rather it will support those people who best express the needs of the time—as happened in the 1970s, when public support swelled for the nongovernmental organizations and politicians who advocated an end to whaling. Here again, the conservation movement can be expected to heed the call. It would be good if fisheries scientists put their hearts in the right place as well.

FISHING MORE AND CATCHING LESS[280]

A BIG PART OF THE SOLUTION

SLOWLY, BUT NOT surely, some of the world's developed countries are extricating themselves from the consequences of the decades-long mismanagement of their marine fisheries, thus solving a part of the global crisis of fisheries diagnosed in the 1990s, characterized by collapsing fish populations and declining industry profits.

The United States, armed with a law—the Magnuson-Stevens Fishery Conservation and Management Act—that mandates rebuilding of overfished fish populations, began relatively early to tackle the twin scourges of subsidy-driven overcapacity (too many boats) and depressed populations (too few fish), and its fisheries are now recovering. In 2013, as a result of pressure from informed consumers, environmental nongovernmental organizations, and fisheries scientists, the European Parliament adopted a series of measures that, if implemented and not watered down, should have similar positive results. These measures include:

- Reducing fishing of European fish populations so that by 2020 they can rebuild each to an abundance that would allow it to generate maximum sustainable yield (MSY), as also required in numerous international treaties and conventions, such as the United Nations Convention on the Law of the Sea of 1982.
- Phasing out the discarding at sea of fish that have been caught without being targeted—a huge problem, often involving 50 percent or more of a boat's catch and consisting of perfectly good fish such as cod and flatfish. Such discarding, often for no good reason, is not only a fisheries management problem but also a moral issue that does not sit well with EU citizens.
- Protecting marine biodiversity, which enables fisheries to exist, and which was taken for granted in the first century of industrial fishing. That exploitation must be accompanied by conservation is a notion understood by a large segment of the European public but not necessarily by European fisheries managers and fleet owners. Now, conservation—including the creation of marine protected areas (MPAS)—will be a part of the reformed Common Fisheries Policy (CFP) of the European Union.
- Regionalizing management, a measure designed to facilitate the application of Europe's CFP to local ecosystems and the participation of local stakeholders. This measure has been applied in the eight Regional Fisheries Management Councils around the United States.
- Increasing the attention given to the "external dimension" of fisheries—that is, to the fact that EU countries have huge fleets operating in the waters of developing countries, notably around Africa and in the Pacific.

Whether these measures will be successful or not will depend on which of the competing groups is more persistent and organized, as it is clear that powerful forces are still arguing that the implementation of, say, the rebuilding mandate and the phase-out of discarding fish would ruin the fishing industry. Such arguments, made from fear—though easily refuted by the example of Norway, a non-EU country that achieved both goals—can be politically effective. Among other things, these arguments have prevented the reform of the CFP to address the issue of the massive government subsidies to EU fisheries, which will contribute to the continued overcapacity of these fisheries. If the reform of the CFP fails, it will probably be for this reason.

WHERE PROBLEMS STILL OCCUR

HOWEVER, IF THE world of fisheries consisted only of the United States and the European Union and their fish populations, one could be relatively Pollyannaish and relax. But both the United States and the European Union are unable to satisfy their seafood demand from domestic catches, well managed or not, so import over 70 percent of the fish and marine invertebrates that they consume—for example, in "all-you-can-eat" shrimp restaurants—about the same percentage as does Japan. These imports are increasing, but the Food and Agriculture Organization of the United Nations (FAO), which assembles and disseminates the fisheries statistics of maritime countries, tells us that since the 1990s the world catch has been either stagnating or slowly declining. So where does the extra seafood for the U.S., EU, and Japanese markets come from?

Most of this seafood comes from the 200-nautical-mile Exclusive Economic Zones (EEZs) of West African countries and from island countries in the Central and South Pacific, where, for example, EU fleets operate in direct competition with fleets from East Asia, notably China, whose leaders also see these countries as larders from which they can get the seafood they want. This is done under an "access agreement" (later renamed "fisheries partnership agreement" by the EU) or illegally, without such an agreement. Yet whether the countries whose fisheries resources are so accessed get a small fraction of their value or not (the going rate is about 5 percent of the value at first sale[281]), the catches taken directly put these countries' food security at risk, especially since the access agreements do not require prior estimation of whether the populations in question—which are usually not managed—are already overfished or not.

The maritime countries of Africa and the small island states of the Pacific have no industrial fisheries; the few industrial vessels that carry their flags are so-called joint ventures, with only nominal links to the host country. In those areas, the foreign fleets compete directly with the local fleets, which are predominantly artisanal, or small-scale, and usually supply local markets. These fisheries are by no means small, however; they supply one-third to one-half of the catch in most developing countries—that is, the part of the catch that is not caught by foreign fleets or exported by joint ventures. These small-scale fisheries thus directly contribute to the food security of rural populations. The work of the Sea Around Us, the research project I lead at the University of British Columbia, demonstrates that this role of artisanal fisheries is widely underestimated, primarily because the developing countries that are members of FAO largely omit small-scale fisheries in their annual reports to FAO,[282] with the result that the estimates of the world catch generated by that organization are far too low and mislead many into believing that industrial fisheries generate most of the seafood that people in developing countries eat.

Fish are tasty and healthful, and thus people should eat fish. But "people" means everyone, including those of us living in Senegal and Mozambique and the Solomon Islands, not only us readers of, or contributors to, the *International New York Times*.

BYCATCH USES IN SOUTHEAST ASIA[283]

DEMERSAL TRAWLING IN SOUTHEAST ASIA

THE BYCATCH OF marine fisheries is a global issue,[284] though it may manifest itself differently, and consequently have different solutions, in different parts of the world. This brief contribution discusses the origins of, and a possible long-term solution to, the bycatch problem in Southeast Asian demersal trawl fisheries, adding to the diversity of mainly North American and European views in the literature. Diversity of views is part of the crane that lifts us to new insights.[285]

Before World War II, several attempts were made to introduce demersal trawling to Southeast Asia, notably by the Dutch in Indonesia, the French in Indochina, and the British in Malaysia. All attempts failed, however, because the gear tested was inappropriate, while the colonial setting was not conducive to industrial fisheries development. Attempts by Japanese vessels were more successful but were never intended to lead to the development of local trawl fisheries and did not do so.[286,287]

The development of indigenous Southeast Asian trawl fisheries started just after World War II in the Philippines and was largely, and literally, driven by landing crafts and other gear and motors left by U.S. armed forces. By the late 1950s, Manila Bay exhibited all the symptoms of what would later be called ecosystem overfishing, and fishing effort began to spill over into other areas of the country.

More important is that the Manila Bay trawl fishery was then being studied by an expert with the Food and Agriculture Organization of the United Nations (FAO), Dr. Klaus Tiews, who subsequently went to Thailand on behalf of a German bilateral development agency. Having seen a Southeast Asian demersal fishery in full swing, he easily convinced the Thai Department of Fisheries to follow suit. An appropriately light, high-opening trawl net was designed and resource surveys were conducted (the first in 1961). The Asian Development Bank provided massive, subsidized credit to would-be investors, and a boom occurred that has now become a fisheries classic, as has the bust that followed.[288]

In the 1970s, Thai trawlers, which had made a clean sweep of the Gulf of Thailand demersal resources, were operating in other Southeast Asian countries, such as Burma (now Myanmar) and Indonesia—sometimes illegally, sometimes not—and once even reached as far as the coast of Oman on the Arabian Peninsula.

More important than these Thai incursions was the adoption of the Thai model of fisheries development by neighboring countries, notably Malaysia and Indonesia. Here, as previously in Thailand, the introduction of demersal trawling led to serious conflicts between trawl operators and small-scale fishers.

TRAWLERS AND THEIR BYCATCH

SOUTHEAST ASIAN TRAWLERS must operate close inshore, for two interrelated reasons: (1) on tropical shelves, demersal fish biomass declines rapidly with depth, far more so than in temperate or boreal shelf ecosystems,[289] and (2) penaeid shrimps, the real target of the demersal trawlers, occur only in shallow waters.

For centuries, Southeast Asian inshore waters have provided a livelihood for thousands of small-scale fishers using a variety of (mainly) stationary gears. Thus, demersal trawlers operating close inshore not only compete for the same resource as the small-scale fishers but also often destroy their passive gears.

Further, Southeast Asian trawlers use material to line their cod end (the narrow end of the trawl net), which has extremely small meshes, usually of less than one inch from knot to knot when stretched. They sometimes cover that with even smaller mosquito netting when aiming for small shrimps or anchovies. Efforts have been made in several countries to increase cod end mesh sizes, but their success has been limited, not only because it is inherently difficult to enforce such regulation, but also because of the nature of Southeast Asian demersal fish, which are mostly small, and the size of penaeid shrimps (and thus require small meshes to be caught).

This tropical region is the world center of marine biodiversity; therefore, these resources consist of an extremely large number of small fish species. The bulk of the biomass is contributed by species not exceeding 6 inches.[290,291] Thus, most fish are as small as the penaeid shrimps, which, because of their high value relative to the fish (around 10:1), are targeted by the trawlers. Given that shrimps contribute about one-tenth of the catch weight, most of that weight will consist of bycatch (the small fish species mentioned above and, to a lesser extent, the juveniles of large species, including snappers, groupers, etc.).

The situation is different with small-scale fishers, who often tend to catch larger fish than trawlers and do not discard part of their catch, as trawlers do. Thus, small-scale fishers are bound to be wary of trawlers, especially when they operate close inshore.

There is much in the literature dealing with the ensuing conflicts, but reviewing that literature is beyond the scope of this essay. Suffice it to say that as these conflicts turned violent, they became destabilizing enough for demersal trawling to be banned in the heavily populated western half of Indonesia[292] and in some areas of the Philippines and to be strictly regulated in Malaysia.

SOUTHEAST ASIAN FISHERIES PRODUCTS

THE CATCH OF small-scale fishers in Southeast Asia tends to be marketed in one of three basic forms:

1. Fresh (iced) or live, as high-quality product (e.g., grouper)
2. Salted and sun-dried, as medium- to low-quality product that can be moved across large distances and that in the 19th century supported a vast international trade[293]
3. Salted and fermented, leading to various "fish sauces," which are added to rice

I will deal briefly with items 2 and 3 because the existence of these traditional products largely shaped perceptions in Southeast Asia about the trawlers' practice of discarding bycatch.

Item 3 is one of the major sources of animal protein in the rural parts of Southeast Asia—even inland. It is a highly nutritious product, beneficial not only for the protein it contains but also as a source of iodine and of calcium, because the (dissolved) bones are also eaten. This product is affordable in a way that larger fish, which tend to be sold whole, are not. Further, many groups in Southeast Asia—for example, the Javanese—simply like small bony fish (e.g., ponyfishes, Family Leiognathidae) and hence do not perceive them as trash fish, even if trawl operators do.

Several of the products in item 3 are poorly described as "sauce." The most important of these products is hard to imagine when one has not seen, smelled, or tasted it. It has the fluidity, and sometimes the color, of olive oil (and is often lighter; hence its name "fish water" in Thai and Vietnamese). It smells "fishy" and has a taste that is mostly salty. It consists of whole fish liquefied by a fermentation process driven by the fishes' own enzymes.[294] (The ancient Romans consumed enormous quantities of a similar product, called *garum*, which was traded throughout the Mediterranean in amphorae—not all were used for wine!—and which seems to have been produced, at least in Mediterranean Spain and southern Italy, until the late Middle Ages.)

The key advantage of this product—called *nam pla* in Thailand, *nuoc mam* in Vietnam, *petis* in Indonesia, *patis* in the Philippines—is that it can help turn large quantities of tiny, and sometimes partly decomposed, fish into a highly esteemed and stable product (a bit like smelly cheese in France). This allows for the use of a seasonal raw material that otherwise would be lost, especially where refrigeration is not available.[295]

The products in items 2 and 3 may be considered Southeast Asian "pre-adaptation" to the emergence of the trawl fisheries and the "trash fish" they create. Trawlers have become the major suppliers of raw material for such products. Efforts are also being made to develop new products; these efforts have been quite successful in some parts of Southeast Asia, but discarding continues elsewhere.[296]

The creation of "trash fish" by the trawl industry occurred at two levels: (1) conceptually—before the emergence of the trawl fisheries, the concept itself did not exist, as all fish that were caught were also consumed, and (2) actually—by reducing the fraction of large fish in the non-penaeid catch and increasing the fraction contributed by their juveniles (one kind of trash fish) and of various, smaller fish (the other major kind of trash fish).

The emergence of trash fish is thus due to the combination of growth overfishing (which removed most of the older representatives of large species, leaving only the juveniles) and of ecosystem overfishing (which saw large K-selected species replaced by smaller r-selected species) that characterized trawling in Southeast Asia.[297]

Marine fisheries resource conflicts in Southeast Asia—particularly conflicts between small-scale fishers and trawl operators—will tend to abate if economic growth continues and human population growth does not. There are already indications of trends in several countries toward returning the exploitation and management of inshore resources to the small-scale fishers and protecting the fishing grounds from trawlers (e.g., by spiking shallow-water areas with strong artificial reefs made of concrete). Because of the selective nature of their gear and of the type of products marketed by small-scale fishers, this return of inshore resources to small-scale fishers would markedly reduce the bycatch problem in Southeast Asia. This scenario might also be a future scenario for other parts of the world.

ON RECONSTRUCTING CATCH TIME SERIES[298]

THE CATCH IN USING CATCH STATISTICS

IT IS WIDELY understood that catch statistics are crucial to fisheries management. However, the catch statistics routinely collected and published in most countries are deficient in numerous ways. This is particularly true of the national data summaries sent by the statistical offices of various Caribbean and Pacific countries to the Food and Agriculture Organization of the United Nations (FAO) for inclusion in its compilation of global statistics.

A frequent response to this situation has been to set up intensive but relatively short-term projects devoted to improving national data reporting systems. The key products of these projects are detailed statistics covering the (few) years of the project. These data are usually hard to interpret, however, given the frequent absence of data from previous periods, from which changes could be evaluated. This is a major drawback, as it is the changes occurring within a long-term dataset that provide the basis from which trends in the status of the resources supporting various fisheries can be determined.

Reconstructing past catches and catch compositions is crucial in order for fisheries scientists and officers in the

Caribbean or the Pacific to fully interpret the data from current data collection projects. This may be illustrated by the following example: Suppose that the Fisheries Department of Country A establishes, after a large and costly sampling project, that its reef fishery generated catches of, say, 5 and 4 tons per square mile for the years 1995 and 1996, respectively. Are these catch figures high values, allowing an extension of the fishery, or low values, indicating an excessive level of fishing?

One approach would be to compare these figures with those of adjacent Countries B and C. However, these countries may lack precise statistics or have fisheries using different gear. Furthermore, A's Minister in Charge of Fisheries may be hesitant to accept conclusions based on comparative studies and require local evidence before making important decisions affecting the local fisheries. One approach to deal with this very legitimate requirement is to reconstruct and analyze time series, covering the years preceding the recent period for which detailed data are available and going as far back in time as possible (i.e., to the year 1950, when the annual FAO statistics begin). With such data, covering the early period of fisheries, it would then be possible to quickly evaluate the status of fisheries and their supporting resources and to evaluate whether further increases of effort would be counterproductive or not.

THE HOW OF CATCH RECONSTRUCTIONS

THE KEY PART of the methodology proposed here is psychological: one must overcome the notion that "no information is available," which is the wrong default setting in dealing with an industry such as fisheries. Fisheries are social activities, bound to throw large "shadows" on the societies in which they are conducted; that is, they interact with other sectors of the economy and are observed by people who are not fishers. Hence, records usually exist that document some aspects of these fisheries. All that is required is to find them and to judiciously interpret the data they contain. Important sources for such an undertaking are the following:

- Old files of the department of fisheries
- Various peer-reviewed publications
- Theses and scientific and travel reports, accessible in departmental or local libraries or branches of the University of the West Indies or the University of the South Pacific or through regional databases, such as the Fisheries Management and Information System (FISMIS; Department of Fisheries, Port of Spain, Trinidad & Tobago), or the Pacific Islands Marine Resources Information System (PIMRIS, University of the South Pacific Library, Suva, Fiji), and so on, or the national departments tasked with collecting census data (to estimate fisher numbers) and economic information (to estimate incomes from fishing)
- Records from harbor masters and other maritime authorities containing information on number of fishing crafts (small boats by type; large boats by length, class, and/or engine power)
- Records from the cooperative or private sectors (companies exporting fisheries products, processing plants, importers of fishing gear, etc.)
- Old aerial photos from geographic surveys (to estimate numbers of boats on beaches and piers)[299]
- Interviews with old fishers[300]

ESTIMATING CATCHES AND CATCH COMPOSITION

ANALYSIS OF THE scattered data obtained from the sources listed in the previous section should be based on the simple notion that catch in weight (Y) is the product of "catch per effort" times "fishing effort."

This implies that one should obtain from these sources estimates of the effort (how many fishers, boats, or trips) of each gear type and multiply by the mean catch/effort for that gear type (e.g., mean catch per year per fisher or mean catch per trip). As the catch/effort of small boats and of individual fishers will differ substantially from that of larger boats, it is best to estimate annual catches by gear or boat type, with the total catch estimates then obtained by summing over all gear or boat types.

Moreover, as catch/effort usually varies with season, the catch for each month of the year should be estimated, and then the estimates should be added up to obtain a total for the year. This should be repeated for every component of the fishery (e.g., small-scale, semi-industrial, and industrial).

Once all quantitative information has been extracted from the available records, linear interpolation can be used to "fill in" the years for which estimates are missing. For example, if one has estimated 1,000 metric tons (t) as the annual reef catch for 1950 and 4,000 t for 1980, then it is legitimate to assume, in the absence of information to the contrary, that the catches were about 2,000 t in 1960 and 3,000 t in 1970. This interpolation procedure may appear too simplistic; however, the alternative is to leave blanks, which later will invariably be interpreted as catches of zero, which is a far worse estimate than interpolated values.

Once catch time series have been established for distinct fisheries (nearshore/reef, shelf, oceanic, etc.), the job is to split these catches into distinct species or species groups. Because comprehensive information on catch composition is usually missing, the job of splitting up catches must be based on fragmentary information, such as the observed catch composition of a few, hopefully representative, fishing units. Still, all available anecdotal information on the catch composition of a fishery (i.e., observed composition of scattered samples) within, say, five years should provide reasonable estimates of mean composition if use is made of the statistical principle that, in the absence of further information on their relative contributions, equal probabilities are assigned to the items jointly contributing to a whole. Thus, a report stating, say, that "catches consisted of groupers, snappers, grunts, and other fish" can be turned into 25 percent groupers, 25 percent snappers, 25 percent grunts, and 25 percent other fish as a reasonable first approximation.[301]

A RECONSTRUCTION EXAMPLE, AND BEYOND

THERE IS OBVIOUSLY more to reconstructing catch time series than outlined above, and some of the available methods are rather sophisticated. They are usually not applied, however, because potential users do not trust themselves to make the bold assumptions required to reconstruct unknown quantities such as historical catches. Only by making such bold assumptions, however, can we obtain the historical catches required for comparisons with recent catch estimates and thus infer key trends in fisheries.

One example may be given here. The FAO catch statistics for Trinidad and Tobago for the years 1950–1959 start at 1,000 t and then gradually increase to 2,000 t in 1959. In the mid-1950s, 500–800 t was contributed by Osteichthyes (bony fish), 300–500 t by "*Scomberomorus maculatus*" (actually Serra Spanish mackerel, *S. brasiliensis*), 100–200 t by *Penaeus* spp. (shrimps), and 0–100 t by Perciformes (presumably reef fishes).

Despite their obvious deficiencies, these and similar data from other Caribbean countries are commonly used to illustrate fisheries trends in the region. Fortunately, it is very easy to improve on such statistics. Thus, one author[302] estimated—based on detailed surveys at the major market (Port of Spain) and a few, quite reasonable assumptions—that the total catch from the island of Trinidad was on the order of 13 million pounds (2,680 t) in 1954–1955—about two times the FAO estimate for both Trinidad and Tobago. Moreover, a source exists[303] that provides details of the small-scale fisheries then existing on Tobago, from which fishing effort and a substantial catch can be estimated, notably of "carite" or cero mackerel (*Scomberomorus regalis*). Further, both of these sources include detailed catch compositions as well, indicating that several of the categories with entries of zero in the FAO statistics (e.g., the clupeoids, or sardine-like fishes) generated substantial catches in the 1950s. Other early sources exist that can be used to corroborate this point.

A GLOBAL, COMMUNITY-DRIVEN CATCH DATABASE[304]

THE ISSUE AT HAND

TO MANAGE THE fisheries in their Exclusive Economic Zones (EEZS), countries need to know their catch. Ideally, their department of fisheries or equivalent agency would know much more than that—the size and productivity of the stocks being exploited, the economics of the fisheries, and so on—but it is essential to know about catch, as the goal of a fishery is to generate and maintain a catch and, if possible, to increase it.

As mentioned in the previous essays, the Food and Agriculture Organization of the United Nations (FAO) does maintain a publicly available database of fisheries statistics, based on submissions by its member states, but it covers only landings (omitting discarded bycatch), doesn't identify the EEZS where coastal landings come from, doesn't present the data by sectors (i.e., industrial, artisanal, subsistence, and recreational), and doesn't estimate the illegal and otherwise unreported and undocumented (IUU) catches usually generated by roving distant-water fleets.

THE SOLUTION

A PUBLICLY ACCESSIBLE database that builds on the FAO statistics but overcomes the deficiencies mentioned above has now been created (at http://www.seaaroundus.org). This database covers the fisheries of all maritime countries and territories of the world, from 1950 to 2015, and will be regularly updated. It is based on historical catch reconstructions by about 300 of my colleagues throughout the world, as well as decade-long support of the Sea Around Us by the Pew Charitable Trusts and the technical wizardry of programmers at Seattle-based Vulcan Inc., which complemented three years' worth of grants from the Paul G. Allen Family Foundation.

The database, which also presents catch-related data and indicators (e.g., ex-vessel values of catches, different types of subsidies received by the fisheries of each country, stock-status plots), allows managers, scientists, students, or ocean activists to find out how much is caught in the EEZ of each country and territory by species or group of species, by sectors, or by catch type (discarded or retained).[305] We can thus acquire an understanding of the fisheries that was impossible to obtain previously and that should help improve fisheries management particularly because, for the first time, the database described here allows the performance of large-scale (industrial) and small-scale (artisanal and subsistence) fisheries to be compared on a global basis.[306]

All information on marine biodiversity in the Sea Around Us database is derived from FishBase (http://www.fishbase.org) for fishes and from SeaLifeBase (http://www.sealifebase.org) for invertebrates. These recognized online encyclopedias are closely linked to http://www.seaaroundus.org, meaning that more information about exploited species may be acquired.

The Sea Around Us database also has a spatial expression—that is, the catch data it contains have been plotted in space using knowledge of the global distribution of exploited fish and invertebrates (from FishBase and SeaLifeBase) and of the fisheries that rely on them. The result is that catch maps can be produced by species or by countries, showing, for example, how fisheries have expanded geographically from 1950 to the present. Finally, the database also includes time series of biomass (the weight of the fish left in the sea) for well over 2,000 populations of fish and invertebrates that are exploited commercially in the different ecosystems of the world's oceans.

Using this database, we could demonstrate that the world marine fisheries catches are about 50 percent higher than suggested by the FAO statistics (which can be viewed as a good thing, since it implies the oceans are more productive than we thought) but have been declining rapidly since 1996,[307] which is an issue that needs to be addressed.

This is just one example of what can be done with our database and website. We hope, moreover, that these tools will be questioned by empowered users and that their feedback will gradually improve both. We also hope that, in the process, our website will become a relied-upon, one-stop, go-to place for information on marine fisheries and will thus contribute, via improved fisheries management, to the incomes and food security of the millions of people who depend on fish.

CATCHES DO REFLECT ABUNDANCE[308]

A STRANGE DEBATE

IN DEVELOPED COUNTRIES such as the United States, Australia, and members of the European Union, many fisheries are monitored by fisheries scientists using expensive "stock assessments." To infer the size of the fish populations being exploited, scientists use the age and size distributions of the fish caught, the results of abundance surveys carried out from research vessels, and information about growth and migration from tag and recapture studies. Yet in about 80 percent of all maritime countries, the only data about the fisheries that are made publicly available are estimates of the weight of fish caught each year. Since 1950, the Food and Agriculture Organization of the United Nations (FAO) has published these catch data (which are gathered by officials in around 200 countries) in the FAO *Yearbook of Fishery and Aquaculture Statistics*.

A debate is raging among fisheries scientists about the wisdom of using catch data to assess the health of fisheries. I agree that catch data should be used with care. But the current dispute is sending a message to policymakers that catch data are of limited use. If countries—especially developing ones—start to devote even fewer resources to collecting catch data, our understanding of fisheries, including their impact on marine ecosystems and their importance for local economies, will suffer.

STOCK-STATUS PLOTS

THE DEBATE ABOUT catch data stems from an analytical approach that was pioneered by the FAO and subsequently developed by others, myself included. In 1996, FAO researchers devised what became known as "stock-status plots."[309] For 400 well-studied fisheries, the researchers plotted catch data over time and used the slope of the graphs to assign the exploited fish populations, or "stocks," to different categories, such as "developing," where catches were increasing, or "senescent," where catches had collapsed. The resulting chart was meant to show at a glance how the fisheries had fared since the 1950s (badly, apparently).

In 2001, the FAO method was modified by fisheries scientist Rainer Froese at the GEOMAR Helmholtz Centre for Ocean Research in Kiel, Germany, and a colleague from the Philippines.[310] The Sea Around Us project then used the modified method to produce stock-status plots for all the fisheries in the world for which catch data were available and made the plots available on our website (http://www.seaaroundus.org). The results revealed a similar trend to that shown by the FAO: the number of collapsed fish populations had steadily increased over the years, and by the mid-1990s, 20 percent of the populations exploited in the 1950s had collapsed.

(We classed them as collapsed if their annual catch had fallen to less than 10 percent of the highest ever recorded.) Unfortunately, it took another ten years, and a misguided claim, for the world to take notice.

In 2006, a group of researchers from various institutions used a stock-status plot to project, among other things, that all exploited fish populations would be collapsed by 2048.[311] Unsurprisingly, this projection, although a small part of the study, triggered an avalanche of alarmist headlines: "Seafood may be gone by 2048," wrote the National Geographic. "The end of fish, in one chart," said the Washington Post.

The weirdly precise 2048 date, with echoes of George Orwell's Nineteen Eighty-Four, was widely derided within the fisheries community. Given the myriad factors that can affect fishing—shifts in policy, rising fuel costs, market crashes, and natural disasters—it is impossible to predict where fisheries will be even ten years from now. But, of the various lines of attack that fisheries scientists have used to discredit the 2006 paper, one charge has since gained momentum and stands to do much more damage to fisheries science and management than the original paper. This is the idea that catch data are not useful for determining the health of fish populations. This is wrong—dangerously so.

THE WEIGHT OF THE EVIDENCE

OVER THE PAST two decades, the amount of fish caught from the world's oceans has declined.[312] Factions of the fisheries community disagree about how to interpret this decline, and they dispute the methods used to assign fish populations to different categories, such as collapsed or underexploited. And it is true that catch size is not just affected by fish abundance—numerous factors, such as a change in management or legislation, can also influence the annual haul of fish. But for the vast majority of species, no signal of this downward trend would even exist without the FAO catch data.

When only catch data are available, fisheries researchers can and should use these data to infer fishery status, at least tentatively.[313,314] Even when "stock assessments" are performed or scientific surveys conducted, the information they provide should always be used in conjunction with any and all available catch data. Take, for example, the Canadian northern cod, which unexpectedly collapsed in Newfoundland and Labrador in the early 1990s, even as stock-assessment experts were using state-of-the-art methods to model its abundance.[315] In the years before the collapse, fishers were using either net traps fastened on the seafloor or trawlers, but because the boats could track shrinking shoals, their catch remained high, even as the trap fishers started bringing in fewer and fewer cod. Stock-assessment experts had monitored only the trawler catches.

Discrediting catch data risks hampering analysis and might also discourage efforts to improve the quality of fisheries statistics worldwide. For the vast majority of species, expert stock assessments can cost from around US$50,000 to millions of dollars per population—especially when research vessels are involved—and so are often not feasible. If resource-starved governments in developing countries come to think that catch data are of limited use, the world will not see more stock assessments; catch data will just stop being collected.

Instead of questioning the usefulness of catch data in assessing fish populations, scientists should urge more governments to collect these data (along with data on fishing effort, the economic value of catches, and fishing costs) and devise cost-effective ways to improve their reliability.

As part of the Sea Around Us project—a collaboration between the University of British Columbia in Vancouver, Canada, and the Pew Charitable Trusts, which aims to monitor the impact of fisheries on marine ecosystems—I am leading a project to evaluate the entire body of FAO catch data collected since 1950. So far, my team has gathered information on fish consumption and the tonnage of fish imported and exported, for instance, to verify the catch data of 180 countries and island territories.[316] Our findings suggest that catches, with the notable exception of domestic catches by China, are under-reported by 100 to 500 percent in many developing countries[317] and by 30 to 50 percent in developed ones.[318]

While fisheries researchers continue the important debate about which fisheries are in decline, why, and to what degree, most fishers worldwide are finding fewer fish in their hauls than their predecessors did. Knowing the tonnage pulled out of the oceans each year is crucial to knowing how to reverse this trend.

THE SHIFTING BASELINE SYNDROME OF FISHERIES[319]

WHERE WE ARE

FISHERIES HAVE BECOME a topic for media with global audiences, but then again, fisheries are a global disaster—one of the few that affect, in very similar fashion, developed countries with well-established administrative and scientific infrastructure, newly industrialized countries, and developing countries. The problems are quickly summarized:

- Heavily subsidized fleets, exceeding by a factor of two or three the numbers required to harvest nominal catches of about 90 million metric tons per year
- Staggering levels of discarded bycatch,[320] a large unrecorded catch that perhaps raises the true global catch to about 150 million metric tons per year,[321] well past most previous estimates of global potential
- The collapse, depletion, or recovery from previous depletion of the overwhelming majority of the over 260 fish populations that are monitored by the Food and Agriculture Organization of the United Nations

THE BASIC IDEA

FISHERIES SCIENCE HAS responded as well as it could to this global disaster by developing methods for estimating targets for management—earlier the fabled maximum sustainable yield (MSY),[322] now the annual total allowable catch (TAC) or individual transferable quotas (ITQS). If these methods are to remain effective, fisheries scientists need to closely follow the behavior of fishers and fleets, but this has not been the case, and our tendency to factor out ecological and evolutionary considerations from our models has increasingly separated us from the biologists studying marine or freshwater organisms and/or communities. There are obviously exceptions to this rule, but I believe that it generally applies, and it can be illustrated by our lack of an explicit model accounting for what may be called the shifting baseline syndrome.

Essentially, this syndrome has arisen because members of each generation of fisheries scientists accept as a baseline the population size and species composition that occurred at the beginning of their careers and use this to evaluate changes. When the next generation starts its career, the populations have further declined, but it is their size at that time that serves as a new baseline. The result obviously is a gradual shift of the baseline, a gradual accommodation of the creeping disappearance of resource species, and inappropriate reference points for evaluating economic losses resulting from overfishing or for identifying targets for rehabilitation measures.

USING HISTORICAL OBSERVATIONS

THESE ARE STRONG claims that I can best illustrate by using analogies. For example, astronomy has been able to test hypotheses against ancient observations (including those found in Sumerian and Chinese records that are thousands of years old) of sunspots, comets, supernovae, and other phenomena that were recorded by ancient cultures. Similarly, oceanography has had, since the days of Commodore F. Maury in the 1850s, protocols for consolidating scattered observations on currents and winds, and later on sea surface temperatures; as a result of the latter, the International Comprehensive Ocean–Atmosphere Data Set (ICOADS) has been extended back to 1870, allowing researchers to infer that, indeed, global warming is occurring.

In contrast, fisheries science does not have formal approaches for dealing with early accounts of "large catches" of now extirpated resources, which are viewed as anecdotes. Yet the grandfather of my colleague Villy Christensen did report being annoyed by the bluefin tuna that entangled themselves in the mackerel nets he was setting in the waters of the Kattegat in the 1920s, for which no market then existed. This observation is as factual as a temperature record and one that should be of relevance to those dealing with bluefin tuna, whose range now excludes much, if not all, of the North Sea.

I could list hundreds of such observations drawn from the historical or anthropological literature and elsewhere, but here it may be more useful to highlight two small fisheries-related studies that have attempted to consolidate these observations and which have led, I believe, to important new insights. In the first, a (female) scientist[323] compiled scattered observations of (male) anthropologists reporting on fishing in the South Pacific and concluded that, despite cultural emphasis on the catching of large fish by men, the gleaning of smaller reef organisms by women and children often accounted for as much catch as the more spectacular activities of the men (even though the women's and children's efforts do not enter official catch statistics). This fact, now widely confirmed by field studies, should lead to a re-evaluation of the fisheries potential of coral reefs.[324]

The authors of the second study[325] used the anecdotes in Farley Mowat's *Sea of Slaughter*[326] to infer that the biomass of fish and other exploitable organisms along the North Atlantic coast of Canada now represents less than 10 percent of the biomass two centuries ago. Some colleagues will find it difficult to accept that the early fishing methods could have had such impact, given their relative inefficiency when compared with our factory ships. However, it must be remembered that the large animals of low fecundity at the top of earlier food webs must have been less resilient to fishing than the survivors that are exploited today. That is, the big changes happened a long time ago, and all that we have as evidence are anecdotes.

Developing frameworks for incorporating earlier knowledge—which is what the anecdotes are—into the present models of fisheries scientists would have the effect not only of adding history to a discipline that has suffered from lack of historical reflection but also of bringing into biodiversity debates an extremely speciose group of vertebrates: the fishes, whose ecology and evolution are as strongly impacted by human activities as the denizens of the tropical and other rain forests that presently occupy center stage in such debates. Frameworks that maximize the use of fisheries history would help us to understand and to overcome—in part at least—the shifting baselines syndrome and hence to evaluate the true social and ecological costs of fisheries.

FURTHER THOUGHTS ON HISTORICAL OBSERVATIONS

FOLLOWING PUBLICATION OF the preceding essay, the temptation was great to follow up on it; this essay presents two loosely (or perhaps not even loosely) connected attempts to build on the ideas therein.

ON BASELINES THAT NEED SHIFTING[327]

A FLURRY OF articles in recent years shows that loss of knowledge about the past may have contributed to an acceptance of other losses, such as declines in biodiversity. I first identified this form of collective amnesia in a 1995 article describing how fisheries biologists assess changes in biomass abundance. Every generation begins its conscious life by assessing the state of the world and society around it and using what it sees as a baseline to evaluate changes that occur subsequently. However, the baselines of previous generations are commonly ignored, and thus the standard by which we assess change also changes. I called this phenomenon "shifting baselines."

Thus, those studying wildlife today might be impressed by the abundance of large wild mammals (bears, wolves, various herbivores) in Alaska, while being unaware that such numbers were at one time common in the Lower 48.[328] Therefore, they might not miss the large animals there and might look askance at efforts to (re)introduce previously abundant species.

The shifting baseline phenomenon has been well documented in marine science, including fisheries research. For example, there are cases where a law[329] may require rebuilding fish populations to the level prevailing, say, twenty years before, although populations were already depleted by then, at least as compared with the level fifty years earlier. It is only by combining the declines noted by successive generations that we can get a full appreciation of the great loss of biodiversity that has occurred in the sea and on land as a result of human activity.

But shifting baselines need not be associated with losses, and forgetting can be a good thing. When people who have suffered under the load of a long, stifling tradition emigrate and thus can distance themselves, both geographically and emotionally, from the ancestral conflicts that in their home countries confined them within balkanized camps, a positive shifting baseline occurs in the generations that follow. Positive shifts in baselines also occur after social change. One example is smoking in enclosed public spaces, which was ubiquitous in the 1960s. At the time, change seemed impossible, and the stranglehold that the tobacco industry had on our legislators seemed unbreakable. Then, somehow, anti-tobacco activism, medical science, and common sense coalesced into an unstoppable force—let's call it the Zeitgeist—which overcame all resistance, first in the United States, then in Europe, including France (France!). Now we look back, and our baseline—and especially that of young people—has so shifted that we don't understand how we ever accepted smoking in tight public places. We have collectively forgotten how it felt (and smelled) and how we could even tolerate it—just as we have collectively forgotten how it was when most people were farmers or, even earlier, hunter-gatherers surrounded by Nature that teemed with a diverse animal and plant life.

Similarly, in our culture, it now seems impossible to even imagine that women and minorities could not vote, attend universities, or become elected politicians. In fact, in the West, the very act of questioning these social advances defines fringe culture, just as denying evolution or climate change defines fringe science.[330] And our baselines have shifted so much that most of us no longer believe in the once powerful notion that there are special people, kings and queens and their broods, who should rule us because a deity said so.

Getting back to Earth: for baselines to shift is not always bad. There are many stupid things that must be forgotten even if they have been the rule for thousands of years. Getting rid of these notions will free our minds and enable us to concentrate on things that matter, including some important things that we should remember.

One of these important things is that what we eat should be healthy. We do not need to eat some of the abject stuff that now passes for food but would certainly not be recognized as such by our ancestors. Other important things that we should remember are that we shouldn't be surrounded by pollutants of various kinds, that we shouldn't allow sprawl to eat up natural landscapes, and that we shouldn't allow out-of-control fisheries to eat up the ocean.

Reversing the present destructive trends induced by large-scale, industrial fisheries—which would be possible under a regime in which fuel energy costs its true price—would lead not only to larger, more plentiful fish for small-scale coastal fisheries to catch but also to a world in which fisheries could coexist with whale watching and other forms of coastal tourism. It would be a world in which people could acquaint themselves with the sea as the wondrous habitat of the many life forms that we may eat as seafood or that we may enjoy just for being there. In other words, we want the bad old things to shift away and the good old things to shift back into focus.

THE HISTORICAL DIMENSION IN RESEARCH[331]

SCIENCE PROCEEDS BY accumulation of knowledge. This truism brings to mind an even process, as could be illustrated by a smoothly ascending curve. Different shapes for the curve result from whether knowledge is thought to be reflected by the amount of information, as measured by the number of new scientific publications, which increase exponentially, or whether knowledge is defined as an increase of the number and scope of theories, which grow more slowly.[332] In either case, an increase of knowledge occurs. This contrasts with what may be called the Standard Social Science Model, where, based on T. Kuhn's classic essay[333] the process of natural science is viewed as dominated by sequences of "paradigm shifts" or by different "discourses," each reflecting mainly the vested interests of an elite group.[334]

Yet, in the natural sciences, we do know more about the Earth since plate tectonics replaced earlier, static representations of global geological processes, and we know more about biology since Darwin's selectionist paradigm replaced its creationist predecessor. In both examples, the new paradigm not only explained more than did its predecessors but also spawned new opportunities and methods of investigation.

Thus, a key criterion for a true advance is that the new model or explanation explains more than its predecessor(s)—that it provides a context for incorporating into a coherent body of knowledge more of the empirical evidence established by previous generations of researchers. Thus, when paradigm shifts do occur, an overall increase of knowledge occurs as well, and the naïve view of science operating in cumulative fashion is vindicated, though perhaps in slightly more complex form. What is required for the cumulative increase of scientific knowledge to break down are crises external to science itself, discussed in the next section.

CRISES AND CHALLENGES

CRISES CAPABLE OF interrupting scientific growth in both developed and developing countries may be caused by unstable funding support. In developed countries, examples of such crises were induced, starting in the early 1970s, by failing support for institutions devoted to taxonomy (mainly museums), once a vibrant area of biological research. As a result, a large fraction of the knowledge held by the last working generation of taxonomists is not being passed on to successors.

In many developing countries, the same period has seen, in relative terms, even sharper declines in funding support, often reducing research institutions to shadows of their former selves—even when acute conflicts such as civil war did not cause scientists to flee their jobs and/or valuable archives and specimens to be burnt or looted.

In some disciplines, notably oceanography and climate research, vast programs of data recovery have been initiated, often triggered by the need for the proper baselines required by global climate simulation models. Here, the cumulative process is restored *post hoc*, to bridge the gaps caused by institutional crises. For example, programs currently exist to recover oceanographic and weather data pertaining to areas held by the Axis powers during World War II (the ultimate institutional crisis) and previously unavailable in global databases.

Similar efforts are exceedingly rare in the biological sciences. Many of my colleagues believe that this situation is due to the complex nature of biological data, compared with the straightforward formats required for oceanographic (mainly temperature and salinity) or meteorological (mainly wind direction and strength, and air pressure) information.

However, one could argue that if there is a will, there will be a way. One way, for example, is to define a minimum format for the key biological information and then make an all-out effort to get that information—although it may be difficult to access, it is there. Here, I think of the example provided by the Species 2000 Initiative, which aims to gather, in a single database, the valid scientific names of all the organisms described since the 10th (1758) edition of Linnaeus' *Systema Naturae* and the references that document these names. Another example is provided by occurrence records, called "bioquads" because they contain the four items (species name, source, date, and locality) required for biodiversity studies.[335] About 10 million bioquads exist in the various museums of the world for fishes alone, and recovering and analyzing them will be a challenge similar to those taken up by oceanographers and climate scientists.

Where some information for each species beyond the original description is available, another way to deal with the challenge of recovering the past is to provide a structure for more detailed information to be captured and standardized, tailored to the features of the type of field survey[336] or of the taxon for which information is to be recovered. A taxon-specific but global approach was taken for FishBase,[337] and we hope that specialists for other groups will follow this example, now shown to work in practice.[338]

Approaches also exist for recovering complex ecological information, notably on the structure of the food webs largely defining aquatic ecosystems. Thus, the present state of a given ecosystem (biomass of its various functional groups, fluxes of matter between producers and first-order consumers, predatory fluxes from first-order to higher-order consumers, etc.) can be represented in standardized fashion using Ecopath models.[339] Then the data recovered from the past can be used to modify the contemporary model so that it will tend to represent an earlier state of the ecosystem—for example, before the biomass of major resource species was reduced by industrial fishing.[340]

This establishes the utility of digitizing, documenting, and analyzing, on a global basis, the aquatic species that have so far been identified, the occurrence records that document their distributions in space and time, and their interactions with other species. Moreover, approaches similar to the

Ecopath software can be easily conceived which would allow the validity of this statement to be extended to terrestrial ecosystems as well.

USING RECOVERED KNOWLEDGE TO PREVENT BASELINE SHIFTS

THE QUESTION THAT now emerges is why we would want to do this. After all, this work covers much of the agenda proposed for the U.S.-based Census of Marine Life, an initiative initially costed at US$10 billion, a rather large sum. However, we may wish to compare this with the sum spent annually by governments to subsidize already overcapitalized fisheries of about $30 billion.[341] As every fisheries economist will confirm, subsidies encourage overfishing and depletion of resources. Thus, all of a sudden, the Census of Marine Life does not look so expensive anymore, at least compared with the support given to forces that are currently contributing to reducing biodiversity and of which fisheries are but a small part.[342]

However, the real reason we want to get proper baselines of the marine life we have now, or once had, is that it is only when the concept of sustainability is based on well-established baselines that it means anything. Otherwise, it is nothing but a feel-good concept.

Without firm rooting in scientific, quantified knowledge of what we now have, or had, we will inevitably experience the "shifting baseline syndrome." As I have described, successive generations of naturalists, ecologists, or even Nature lovers use the state of the environment at the beginning of their conscious interactions with it as *the* reference point, which then shifts as successive generations degrade that same environment. The story of the frog kept in water that is heated very slowly comes to mind here, and if we are not careful, we are going to get boiled as the frog does: a runaway greenhouse effect would do the job nicely.

CONSILIENCE IN RESEARCH[343]

CONSILIENCE: DEFINITION AND EXAMPLES

SCIENTIFIC BREAKTHROUGHS HAVE usually been the result of convergence among traditionally distinct disciplines.[344] These disciplines usually arrange themselves in a hierarchy, with logic/mathematics (and the usual criterion of parsimony, or the choice of the simplest scientific explanation that fits the evidence[345]) providing the backbone of any given advance, physics and chemistry providing the basic rules constraining the changes of its material substrate, and evolutionary biology providing the framework that constrains its living organisms (if any), including humans and their culture. E.O. Wilson called consilience[346] (from "jumping together") the explicit search for scientific explanations within the context of this hierarchy and provided examples of research issues whose resolution, he thought, would occur faster if consilience were used as an explicit criterion—in addition to parsimony—for structuring research programs.

My favorite examples (not Wilson's) of consilience are the Cretaceous-Tertiary (K-T) extinction of 65 million years ago, of dinosaur fame, and the origins of *Homo sapiens*, both of which made mutually compatible data and concepts from an enormous range of disciplines that previously had not interacted with each other. A nuclear scientist (Luis Alvarez), his geologist son (Walter Alvarez), and two chemist colleagues proposed that the K–T extinction was caused by the impact of a large meteorite.[347] This hypothesis, then based mainly on evidence from an Italian dig, was subsequently corroborated by petroleum geologists, who had previously identified and then ignored the Chicxulub impact crater, in Yucatan, Mexico. Other scientists, from astronomers to evolutionary biologists, joined the fray,[348] and gradually, the Alvarez hypothesis was accepted by the best of them. The results of this development have been extraordinarily fruitful, "provoking new observations that no one had thought of making under old views."[349] This led to, among other things, a resolution, in evolutionary biology, of the ancient but still acrimonious debate between catastrophists (Sedgwick, Cuvier, et al.) and uniformitarianists[350] (Lyell, Darwin, et al.). Even popular culture was affected (see the movies *Meteor*, *Armageddon*, and *Deep Impact*).

The eventual creation of an international system for tracking potentially dangerous meteorites is not unlikely.[351] Similarly, the paleontological evidence for a recent African origin of *Homo sapiens*, with subsequent dispersal to West Asia, Eurasia, Australia, the Americas, and finally Oceania is supported by archaeological and genetic evidence, with linguistics providing the clincher: an evolutionary tree that closely matches that generated by the physical disciplines.[352] The latter example shows that the hierarchy of sciences implied in consilience does not mean that the specific results of a more fundamental discipline are inherently more reliable than those of a derived, or less fundamental discipline. Rather, it only implies that the different sets of results must be mutually compatible. Thus, the linguistic evidence is, in this example, no less important than the evidence based on genetics. Similarly, when the physicist Lord Kelvin pronounced that the Earth was only a few thousand years old, based on the time required for a large sphere of hot iron to cool off (i.e., based on physics), and that evolution by slow natural selection was thus impossible, it was he who was wrong, not Charles Darwin.[353]

MAKING CONSILIENCE WORK

THE QUESTION NOW is whether consilience can help in the work of oceanographers, marine biologists, and fisheries scientists—that is, in fields perhaps less glamorous than those in the above examples. Various concepts we may call consilient come to mind here. An example is the mass-balance concept—the notion that, in a given system, mass must be conserved, irrespective of its movements and transformations. This principle is related to the first law of thermodynamics, which states that energy can be neither destroyed nor created. For chemical reactions, this implies, among other things, that "the sum of the masses of the reactants must equal the sum of the masses of the products."[354] Physical oceanographers also rely on mass-balance when calculating geostrophic flows from density fields[355] or when calculating upwelling intensity from coastal wind stress, which implies water masses welling up to replace water blown off the coast.[356]

On the other hand, one rarely hears biologists, or even ecosystem modelers, explicitly invoke the principle of mass–balance, though it is also an absolute requirement for living things.[357] One exception to this is the ecosystem modeling work of T. Laevastu and colleagues, in which mass–balance was used as a key structuring element for trophic interactions and migrations.[358,359] J.J. Polovina emphasized this feature when he simplified Laevastu's model and formulated the Ecopath approach,[360,361] thus giving it the feature that made it applicable to a wide range of system types.[362,363] Ecopath uses the mass–balance approach to verify that the estimated production of the functional groups (exploited or not) of a given ecosystem matches the estimated consumption by their predators. Such verification is not ensured by the publication of individual estimates, however precise, even in the best journals catering to the different subdisciplines of marine biology.[364]

Rather, it is by incorporating such estimates into a mass–balance ecosystem model that we verify their mutual compatibility and hence assure ourselves of their reliability and usefulness. This should have a beneficial impact on marine biology, whose work on different processes in an extremely wide variety of organisms is sometimes perceived as lacking cohesion. Moreover, ecosystem and mass–balance considerations should help renew fisheries science as well, given that it has been too narrowly focused on the study of single species and on industrial fisheries, and usually overlooks bycatch-discarding practices, non-commercial species, and other fisheries (artisanal, recreational, etc.) as well.

The relation of these points to Wilsonian consilience should be obvious. Consilience also implies developing protocols for integrating the results of (satellite-based) remote-sensing studies[365,366] into mass-balance trophic models of ecosystems. The key results relevant here are (1) definition of geographic system boundaries, as in the case of A.R. Longhurst's "biochemical provinces,"[367] and (2) estimates of productivity of the planktonic algae (i.e., the primary production) of the different oceanic provinces.[368] Here, by constraining model size, remote sensing can link with ecosystem modeling and thus work in consilient mode. Note also that both remote sensing and trophic modeling may be accused of being "superficial"—remote sensing because, quite literally, it cannot look deeper than a foot into the sea, and trophic modeling because it does not consider interactions other than those generated by plankton grazing and predation. Yet, when data from the two approaches are analyzed jointly, inferences can be drawn that go well beyond those based on more conventional approaches.[369,370]

Perhaps we may infer from this issue of apparent superficiality that consilient work may suffer, at least in some cases, from analogies with multidisciplinary work, in which the methods of different disciplines are brought to bear on a given topic (e.g., as chapters in a book), without relating any of these methods to each other. Such unconnected work is all too frequent—for example, in that discipline called coastal area management. Consilience, it seems to me, should also apply to the strengthening of inferences that results from the past being related to the present. This is what occurs when we use knowledge gathered by historians, or by scientists of past centuries, and often perceived as anecdotal, to establish stable baselines for biodiversity.[371,372]

This is also what occurs when we draw inferences from time series, whose increasing length increases their contrast and hence their usefulness for various analyses.[373] Thus, it is important that the physical and biological time series generated by various scientific programs be continued, as they become more valuable the longer they get.

The last aspect of consilience to be covered here is its implications for the languages we use. Trivially, this means that we must speak the same vernacular language (in science, it is English, as it mostly turns out), and, only a bit less trivially, we must translate concepts in and out of our various discipline-specific jargons. Also, and this is where things start getting really complicated, we must identify concepts that cut across disciplines and a multidisciplinary currency allowing for transactions between disciplines.

I shall illustrate this with reference to coastal area management (CAM), a discipline that has many practitioners and applications but whose defining tenet(s), topic(s), and technique(s) remain elusive. Some practitioners give the impression that anything that happens anywhere on or close to any coastline is within the purview of CAM and that any method ever used to investigate any of these things is appropriate for CAM. (The reader will understand that I could not provide references to back these claims without antagonizing people whose technical work I respect in spite of the disorganization of their discipline.) Thus, for researchers on CAM, as everywhere in science,[374] the challenge is to identify tractable problems related to its defining objects: coastlines.

Coastlines differ from other geographic features in that most of their different characteristics are arrayed in a single dimension—that is, in the form of transects that are perpendicular to the coastline and that stretch from upland to the sea. In contrast, fewer differences occur parallel to a coastline. Alexander von Humboldt, a founder of physical geography, was the first to use transects to document geographical variations along strong gradients.[375] Transects have also been introduced to agroecosystem analysis,[376] on the other hand, as a tool to express in a simplified manner the complex interactions within highly integrated farming systems.[377] From this, it is straightforward to propose that multisectoral coastal transects should become a key concept in CAM and that, suitably formalized, such transects could lead to the common currency required for comparison of coastal systems, and for comparative evaluation of various injuries to such systems.[378]

The SimCoast software, which was used in the EU Gulf of Guinea Project, implements this coastal transect approach and provides, via fuzzy logic, the currency that CAM had been lacking so far, enabling quantitative comparisons of impacts due to different, otherwise incommensurable agents,[379] ranging from upland erosion to fisheries policy. That this should lead to consilience among, and progress for, the different disciplines so far fruitlessly engaged in CAM needs little emphasis.[380]

Another example of consilience requiring a common currency is FishBase, the electronic encyclopedia of fishes (see http://www.fishbase.org), which works only because a standardized nomenclature[381] is used to establish the links between the widely different data types included in this database.[382] This is what enables FishBase to provide, among other things, a comprehensive coverage of the fishes of any region of the world and of their biology.[383]

The ventures given as examples here show that scientists from a variety of countries, both developed and developing, can contribute to a common project. And this is probably the neatest thing about consilience: it implies that we can all contribute, given some self-discipline.

FOCUSING ONE'S MICROSCOPE[384]

THE DISCOVERY OF CELLS

SCIENTIFIC DISCOVERIES ARE often the result of focusing one's microscope—actual or metaphorical—and so rules have emerged on how to focus. For instance, although plant and animal cells were discovered in the second half of the 17th century by Robert Hooke and Anton van Leeuwenhoek (Kaspar Friedrich Wolff, who established that all plants were composed of cells, could be mentioned here, as could Theodor Schwann, who demonstrated the same for animals), it took over 150 years of arduous work to fully establish that cells are the building blocks of all living things.

Establishing this fundamental role of cells was complicated by the wide difference of cell sizes and other properties in various organs and by the existence of acellular tissues in both plants and animals. But throughout these 150 or so years of sometimes bitter debate about the roles of cells, one thing was clear all along: those who didn't believe in cells had to adjust their microscopes in the manner of those who saw them, and not the other way around. Why? Because if the "cell denialists" (for want of a better term) focused their microscopes such that they showed only objects larger or smaller than cells, then obviously cells were not detected. This also was the case if the dye they used to highlight their tissue samples did not generate sufficient contrast, or if they were clumsy and their samples were too thin or too thick for cells to be visible.

In other words: It was easy not to see cells, and this is why we celebrate those who did, along with those other scientists who discovered things that others couldn't see, including natural selection, plate tectonics, and the structure of DNA.

FISHING DOWN MARINE FOOD WEBS

IN 1998, MY coauthors and I first described the phenomenon now known as "fishing down marine food webs," mainly because we were lucky both with the data available at the time and with the setting of our conceptual "microscopes." The catch data then at our disposal pertained to the nineteen large statistical areas that are used by the Food and Agriculture Organization of the United Nations (FAO) to report global catch statistics, and we could detect a strong fishing-down signal in about half of them. We suggested that the fishing-down process might be widespread, but we didn't have a solid explanation at that time for why it did not seem to be occurring in all areas.

"Fishing down" is essentially what happens when the fishes (and invertebrates) of a given ecosystem become vulnerable to fishing—for example, as a result of newly introduced trawlers. In such cases, the larger, longer-lived fishes at the top of the food web (which have high trophic levels) are depleted faster than the smaller, shorter-lived fish and invertebrates (which tend to have lower trophic levels). Thus, time series of multispecies catches from the ecosystem and assemblage in question will exhibit declining mean trophic levels.

Subsequent research that my associates and I undertook, as well as research by a number of independent authors throughout the world, has helped to address the arguments of early critics of the fishing-down concept and to establish its general occurrence (see http://www.fishingdown.org) and intensity (0.05–0.10 trophic levels per decade). We were also able to identify many of the factors that can cause the effect to be masked, thus knocking our microscope out of focus.

I have reviewed much of this work—in chapter 2 of a book called 5 *Easy Pieces: The Impact of Fisheries on Marine Ecosystems*[385]—with reference to so-called "judo arguments," Isaac Asimov's name for points that your opponent makes, but which can be turned around and actually strengthen your case. For example, one judo argument against the fishing-down concept was that it originally considered shifts in between-species composition but not within-species changes in size and hence trophic level. This particular point turns into a judo argument once you consider that, when fishing intensifies, large fish (e.g., cod, grouper, or tuna) become smaller and hence tend to have lower trophic levels, trends that intensify the fishing-down effect. The other judo arguments were similar, with the fishing-down denialists repeatedly ending on the mat.

MASKING FACTORS AND THEIR EFFECTS

IN 2010, NATURE published a paper on trends in fisheries[386] that also had its microscope out of focus and that consequently presented a confused picture, with fishing down sometimes visible, sometimes not. The equally confused prose of that paper's authors included several judo arguments; notably, they did not consider the spatial expansion of fisheries, which is one of the strongest masking effects for fishing down. Proceeding at rates between 0.4 million and 1.5 million square miles per year from 1950 to the beginning of the 21st century,[387] fisheries expansion is a masking factor that had already been identified and warned against. If you exploit a shelf ecosystem with a trawl fishery that reduces the abundance and size of the big fish and, in fact, the biomass of the entire assemblage of exploited species, you will eventually need, if you want to maintain your catches, to expand into deeper, offshore waters to access previously unexploited, large high-trophic-level fishes.[388] A beautiful example is provided by Chinese fisheries in the East China Sea, where the mean tropic level of the coastal fishery declined from 4.0 in 1978 to 3.5 in the 21st century, while the trophic level of the offshore fishery, initiated in the 1980s, declined from 4.3 in 1990 to 4.1 in the 21st century.[389]

Consequently, if you compute time series of mean trophic levels in catches from an expanding area and do not account for the expansion, chances are that you will fail to detect any fishing-down effect. Masking effects of this sort are the reasons why, in science, we standardize key variables. For example, agronomists working on increasing rice productivity use standardized paddies for comparing different treatments (e.g., with and without a certain type of fertilizer), but do not increase the area planted.

Similarly, when making statements about the health of the global ocean or the status of the world's marine fisheries, researchers must use studies that do not represent a grossly biased sample, drawn from the well-managed fisheries of a few countries or regions at the world's end, like Alaska or New Zealand, lest one's microscope be, again, out of focus.

HOMO SAPIENS: CANCER OR PARASITE?[390]

ARE WE "PART OF THE ECOSYSTEM"?

AS A FISHERIES scientist, I am usually expected to reflect on the destruction of life in the ocean that goes along with the industrialization of fisheries. I have done so in numerous publications[391,392,393,394] and in several essays in this book. This essay, however, will not be about what a detractor called the litany. Rather, it is an attempt to connect two more fundamental memes, both of which deserve far more scrutiny than they have received so far. They are (1) the notion that we humans are "part of the ecosystem," frequently asserted by those who try to reconcile human exploitation and the maintenance of the ecosystems in which the exploitation occurs,[395,396] and (2) the notion that we humans are "a cancer on the Earth," proposed by less Pollyannaish authors.[397,398]

The first of these notions, that we are "part of the ecosystem," is treated as a truism in the fisheries and marine conservation literature[399,400] and is reflexively invoked to dismiss schemes that propose setting up Nature reserves from which all extractive activities by humans are excluded.[401]

This hopeful notion is obviously suffused with good intentions, as we would like to reconcile Nature with human well-being whenever possible, just as we would like to have our cake and eat it, too. But if taken seriously, this notion prevents us from thinking of ecosystems without human activities in them. For example, we ought to be able to at least conceive of a natural park in which one cannot mine, fish, hunt, or drive snowmobiles. Useful concepts should help us to think about potential scenarios, not preclude them because of our own definition.

Obviously, there was a time when we were indeed very much part of our ecosystem. Thus, our various ancestors in the African savanna, while chasing after antelopes armed only with pointed sticks, could themselves easily fall prey to another predator, such as a lion. In fact, in those times, the population of humans was largely controlled by the dynamics of their predators, along with the dynamics of their food supply. In effect, our population was then controlled both from the top down and from the bottom up, with the result that our ancestors' population was not able to grow beyond the carrying capacity of the African savanna.[402,403]

At some point, though, we acquired one or several traits through natural selection or cultural evolution that allowed us to escape control by large predators, probably via a mixture of reciprocal altruism and language,[404] enabling collective defense (and offense). As a result, we could eliminate our carnivorous competitors and our population could grow so as to exceed the carrying capacity of the environment for hunter-gatherers.[405] One consequence was murderous conflicts over resources, something that has accompanied us ever since, although perhaps with diminishing intensity.[406]

Another consequence was our expansion out of Africa into the rest of the world.[407,408,409]

This expansion was performed by hunter-gatherers, except for the more recent expansion into Oceania, where agriculturists colonized one island after the other.[410] Whether by hunter-gatherers or agriculturists who also hunted, the expansion always had the same results: the elimination of potential predators (most large carnivores) and of large prey—such as mammoths, mastodons, giant ground sloths, and horses in North America[411] and moas in New Zealand.[412] Another common result was the degradation of the vegetative cover, due to the absence of the cropping and fertilization by large herbivores,[413] fires,[414] and agriculture-induced erosion.[415]

The invention of agriculture made us less dependent on the fluctuating abundance and migration cycles of prey animals and on knowledge of their habits.[416] We also modified the plants that we found in various habitats and adapted them to our requirements.[417] Here again, this development can be interpreted as removing us even more from natural ecosystems and cycles, since the plants we farmed, while originally wild, became part of alternative ecosystems and cycles meant to support us and only us. Industrialization, with its use of fossil energy to produce fertilizers, and the discovery of the germ theory of disease (which led to public sanitation systems, to improved personal hygiene, and later to antibiotics) were further steps out of ecosystem control into cycles meant only for us. These cycles, which eventually came to be called the economy, operate within—but are not part of—natural ecosystems, with which, however, they interact in multiple, pernicious ways,[418] leaving a vanishingly small fraction of Nature not grossly disrupted by the human enterprise.

ARE WE A CANCER OF THE EARTH?

HUMANS HAVE PROGRESSED from being an undisputable part of Nature to destroying it through their economy.[419] This progression can be seen as a frightening analog of the progression of cancer tumors in someone's body, where Nature is the "body"[420] and individual species are its cell types, with multiple sets of controls previously ensuring homeostasis. (Yes, things change greatly in geological time, too, but never as fast as we impose on the Earth—except for a meteor slamming into the Earth, another analog to our ecological impact, but not one pursued here.)

One of the "cell types," *Homo sapiens* in this case, has managed through crucial changes—for example, the invention of language[421] or collective hunting[422] or pointier weapons—to escape these controls and proliferate, using the other "cell types" (our domesticated plants and animals) as substrates. The literature on humans as cancer of the Earth provides very detailed analogies (or is it homologies?) between the growth of civilizations and the global economy, and the growth of cancerous tumors. In fact, the close match between these two groups of phenomena is frightening, just like the realization when flying over any landscape that whatever human impact we notice out of the plane's window is something that has grown and will continue to grow—until when?

Most cancer cells are stupid in that they kill their host once they have lost all the genes that, by constraining their multiplication, compelled them to function as part of organized tissues. One of the few exceptions here is transmissible cancers such as the one that is causing the Tasmanian devil population to plummet.[423]

Parasites are evolutionarily smarter. They may be very infectious at first, but usually a strain will select out that can co-exist with the host and may in fact turn into a symbiont,[424] such as the benign bacteria that protect us from other, potentially harmful bacteria.[425]

OUR EVER-EXPANDING ECONOMY

CANCER CELLS, IN contrast to parasites, become ever more virulent as a cancer progresses. Our economy is becoming more virulent as well. For centuries, it was fueled by "normal" returns on capital, from about 5 percent[426] to 10 percent per year.[427] The economy driven by normal profit consisted of "real" processes, such as manufacturing goods or transporting them from sites of production to markets. However, an increasing segment of our economies is not distinguishable from a set of interacting Ponzi schemes,[428] characterized by huge discount rates[429] and short-term profits, negating the very idea of sustainability. This "Ponzification," also known as the "Wall Street-ization" of the economy, implies that a firm capable of generating a 5 to 10 percent return in the long term will likely be eaten up by a financial institution seeking super-profits of 20 percent or more per year in the shorter term. Natural wealth-generating processes, such as the (re)growth of forests or the growth of wild or domestic animal populations, do not live up to these expectations,[430] which can be met, therefore (and only for a while), by liquidating assets or through Ponzi schemes such as the one perpetrated by Bernie Madoff.[431] Hence the clear-cutting of forests worldwide,[432] the decimation of fish populations,[433] and the bankruptcies (with subsequent asset-stripping) of previously profitable private and public enterprises that otherwise do not generate the super-profits sought by financial banks and hedge fund managers. This scenario leaves few public resources to address structural problems, both within countries (health, education, infrastructure) and between countries (development issues, global warming). Wars may thus continue to plague us, including wars in which nuclear weapons are used,[434] one of which may be terminal.

COULD WE BE BENIGN PARASITES?

THE QUESTION IS, therefore, whether it will be possible to turn us humans into benign parasites on the surface of the Earth, whose various evolved ecosystems would retain their ability to function, or whether we will continue to be part of the Earth's ecosystem in the same way that a malignant tumor is—never for a long time—part of a person's body: All bets are off.

ACADEMICS IN PUBLIC POLICY DEBATES[435]

A CANADIAN VIEW OF A GLOBAL ISSUE

ACADEMIC TENURE IS a difficult topic, and while it is easy to get on one's high horse and claim that it is a vital element of higher education, it may also be that for some academics tenure encourages sloth. This essay, however, focuses on the role of academic tenure at research universities in a country I know well and on its potential role in maintaining the integrity of the environmental sciences.

Canada, despite its long democratic tradition, has a record of attempts to suppress inconvenient scientific findings. The well-known Olivieri Affair revealed the extent of the ethical swamp within which university leaders—especially in medicine—operate to protect lucrative associations with the private sector, specifically, the pharmaceutical industry.[436] Here, however, the focus is on the plight of the beleaguered scientists in Canadian government laboratories.

From 2006 until 2015, Canada was benighted by a conservative government that magnified a pre-existing tendency for the heads of government agencies and laboratories to prevent "their" scientists from speaking up about issues in their areas of expertise.[437] This tendency was vividly illustrated to me, then new to Canada, at a public debate at the University of Ottawa by a former high-ranking official of the Canadian Department of Fisheries and Oceans (DFO), who asserted that DFO staff owes loyalty to the Queen (this is Canada!) and thus to the Queen's Minister, and not to the citizenry. This event occurred only a few years after the 1992 collapse of the northern cod fishery, which put 50,000 people out of a job (but not the Queen) and required immense amounts of taxpayers' money to mitigate.

EARLIER CASES OF MUZZLING SCIENTISTS

RIGHT AFTER THAT collapse, a well-known DFO scientist who had the audacity to argue that the collapse of northern cod was not due to abnormally cold temperatures and not to hungry seals (the perennial villains in Canada)[438] but to government-sanctioned overfishing was officially reprimanded for speaking up, although it is now well established that he was right[439] and that fisheries management in Canada is below par.[440] R.A. Myers, the scientist in question, then took refuge at a university, where he became a tenured faculty member and played a critical role in convincing the world that northern cod had not been the only formerly abundant fish population reduced by overfishing to a shadow of its former self.[441]

Although it is legitimate for governments everywhere to expect restraint from civil servants, the conditions on Canadian government scientists, and particularly those working on environmental issues, were so restrictive[442]— at least in comparison with those of other Western democracies—that they became a topic in respected international scientific outlets,[443,444] in a book titled *War on Science*,[445] and in numerous articles and editorials in Canadian media.[446,447,448]

A VIRUS IN FARMED SALMON AND "ETHICAL OIL"

THE ABSURDLY HIGH level of pressure exerted on government scientists may be illustrated by the events following the discovery of a viral signature in (wild) sockeye salmon[449] already threatened by "sea lice," parasites emanating from farming operations relying on introduced Atlantic salmon,[450] which were also the likely source of the virus in question, via infected eggs imported from Norway, where the virus is common.[451] DFO is mandated to encourage this risky form of mariculture, and thus the first author of the paper in question, a DFO staff member, was not permitted to talk about her discovery publicly, under the pretext that she would later testify at the Cohen Commission, set up to investigate the decline of wild Pacific salmon in British Columbia. Her eventual (filmed) deposition, consisting mostly of monosyllabic answers, was typical of what occurs when people are afraid.[452] Such a degrading situation should not occur in science and certainly not in democracies.

Since the ascent of a government that, after its successful renaming of tar sands to oil sands, attempted to rename the muck extracted from Canadian *tar* sands ethical oil (because it originates in a country where women can drive cars, as opposed to Saudi Arabia, where until recently women were not permitted to do so), stories such as this have been common. Government scientists were able to speak to the press only with government minders—21st century political commissars—present. Entire laboratories specializing in ecotoxicology and Arctic ecology were closed, so no one would be left to study the effects of exploiting the Canadian tar sands and drilling for oil in the high Arctic on the health of humans and ecosystems. The Fisheries Act was defanged when the protection of freshwater fish and their habitats was lifted,[453,454] and so oil development could proceed without hindrances such as laws protecting the environment. Obviously, the Canadian cabinet was, with regard to global warming, firmly in the denialist camp, despite the absurdity and destructiveness of this position,[455,456] which diminished Canada's standing in the world community.

TENURED ACADEMICS MUST SPEAK UP!

UNDER THESE CIRCUMSTANCES, who could speak for science in Canada? To the extent that the autonomy of universities was still respected and the tenure system still worked, there was at least one group of scientists in Canada who could object to the silencing of scientists and to what appeared to be preparing the ground for turning the country into a petro-state. One might argue that because tenured faculty can express their findings and views with relative impunity, they have a duty to do so when their colleagues in governments are being muzzled. By extension, scientific organizations such as the Academy of Science (part of the Royal Society of Canada) and the Canadian Association of University Teachers, which are composed of mostly tenured academics, can bring egregious breaches of scientific integrity to the attention of the media and the public.

Such groups—which, for environmental science, include the Canadian Society for Ecology and Evolution, the Society of Canadian Limnologists, and the Canadian Society of Zoologists, among others—have communicated with the media about the matters discussed here. The societies provide an important vehicle for nontenured scientists to add their voices to the debates and discussions without risking retaliation.

Generalizing, one can also note that academia, as shaped by tenured faculty, is one of the few sectors in Canada (and even more so in the United States) that is not in the hands of corporations—though the increasing privatization of the higher education sector, and reliance on nontenured or sessional lecturers,[457] is gradually undermining this bulwark as well. Similar considerations apply in many other parts of the world and may also be relevant when discussing the tenure system in education.

A HAPPY POSTSCRIPT

ON OCTOBER 19, 2015, the Liberal Party won a resounding electoral victory, and the Conservatives were swept from the government they had controlled since 2006.

Among the many reforms that the new prime minister, Justin Trudeau, immediately began to implement was a directive to "un-muzzle" scientists working for the Government of Canada.[458] Shortly thereafter, these scientists "successfully negotiated a clause in their new contract that guarantees their right to speak to the public and the media about science and their research, without needing approval from their managers."[459]

WORRYING ABOUT WHALES[460]

FISHERIES AND THE MARTIANS

ON MAY 8 and 9, 2008, I had the opportunity to attend a workshop in Dakar, Senegal, organized by the World Wide Fund for Nature (WWF) and the Lenfest Ocean Program (LOP) and devoted to the interaction between the great whales and the fisheries of Northwest Africa. The workshop was titled "Whales and fish interactions: Are great whales a threat to fisheries?" and was attended by officials from the fisheries ministries of half a dozen countries in the region, from Mauritania to Guinea; WWF and LOP staff; a few scientists; and, most interestingly, parliamentarians from the host country.

Baleen whales come to that part of the world to reproduce, and there are no live observations or stomach content analyses indicating that they actively feed during that time (even from several decades ago, when there was some occasional whaling off Northwest Africa). This is in line with what is known about baleen whales elsewhere in the tropics. When baleen whales feed, they rely mostly on krill and other small planktonic organisms, and thus they would not, in any case, interact with the demersal and tuna fisheries prevailing off Northwest Africa. So, why hold a workshop on this outlandish topic? Why not Fisheries *vs.* the Martians?

The reason for the workshop was not only that the countries in the Northwest African region were increasingly voting with Japan at meetings of the International Whaling Commission but also that their delegates justified such votes on the grounds that their fisheries were negatively affected by baleen whales. In fact, they argued that the whole ecosystem is "out of balance," a balance that could be re-established only by killing whales. Never mind that this idea flew in the face of everything known about the fisheries of the region, whale biology, and common sense. And the idea did not improve when it was tailored for local consumption.

This was a very awkward situation for me to be in. I have worked for years on West African fisheries with colleagues from the region and have supported their countries' justification of the activity of European Union–based or other distant-water fleets operating in West Africa on the basis of questionable "agreements," many of which the coastal countries were blackmailed into signing and through which their fisheries resources were made available at less than bargain prices.[461] These distant-water fleets—jointly with the local, totally unmanaged, and overgrown "small-scale fisheries"—have reduced the fisheries resources off West Africa to shadows of their former selves, making management of these fisheries, and especially a reduction of their aggregate effort, a priority.

This, in fact, was the main result of the EU-funded international research project called "Système d'Information et d'Analyse des Pêches de l'Afrique du Nord-Ouest" (SIAP).

This project, which I helped design, provided for West African scientists and others to collaborate on the analysis of over half a century's worth of catch time series and other data. The results were presented at an international conference held in Dakar in 2002,[462] amid a flurry of articles in the local press.

WHAT WE KNOW

THIS WAS NOT the first time, obviously, that such findings were reported. In fact, the SIAP project was largely based on gathering and analyzing the vast literature, spanning several decades, that tracked the declining trajectory of the fisheries off West Africa. This literature, and the syntheses that resulted from the SIAP project, are available to inform local policymakers interested in reforming fisheries policies. The most crucial reform would be moving from a situation where West African waters are seen as a larder from which an endless supply of fish can be extracted to supply foreign markets[463] to one where West African countries could build on the export and processing of fish to strengthen their own economy and benefit their own people.

The government positions that I heard at the May 2008 meeting suggested, however, that such reforms were not being contemplated. Instead, the top fisheries officials of West African countries appeared to have thrown in their lot with their Japanese advisers and their whales-eat-our-fish mantra, for reasons that are either obscure or too obvious to mention.

The excellent scientific presentations at the workshop by Drs. Kristin Kaschner and Lyne Morissette[464] dealt with the biology of the baleen whales off Northwest Africa, their behavior, their incorporation into (Ecopath) trophic models, and the results of some preliminary simulations (with Ecosim), which suggested that killing all the whales off Northwest Africa—even if it could be done—would have little effect on the fisheries resources and catches.[465]

At every step, their findings and assumptions were questioned by one or the other government official, using concepts (such as "ecosystem balance") and arguments ("You have not studied the stomachs of newborn calves off West Africa, so you don't really know that they don't eat our fish") originating in the Tokyo-based Institute of Cetacean Research. The only evidence these officials presented was evidence of bad faith, the whole line of arguments being based on absent data. These purely negative arguments are of the same kind as those that advocates of the so-called "intelligent design" use to criticize evolution by natural selection, but (for good reasons) never offer a positive argument for the case they attempt to make.

THE MAYOR TO THE RESCUE

THERE WAS A ray of hope, though. The participating Senegalese parliamentarians, both from the Senate and the House, were united in their questioning of their government's position and in mentioning their surprise at a government policy that has never been publicly debated and that is actually alien to the culture of their constituents.

This very point was emphasized by a parliamentarian and mayor of a fishing town, who mentioned that her constituents, far from considering whales to be their competitors, consider them their guardians and want to see them protected. This view was echoed by participants from other West African countries.

Still, I left Dakar with a heavy heart. To see that such a great country as Japan has twisted its entire development aid and corrupted fisheries officials of an entire region for the sake of its tiny, heavily subsidized whaling industry is sad. It will probably be years before the countries targeted by these delusional policies will see through these maneuvers and free themselves from the officials who mislead them. Also, the real potential of whale ecotourism is not being explored, although it has become a serious source of foreign currency in various other countries—for example, Argentina.[466]

Foremost, however, the countries successfully targeted by the whales-eat-our-fish delusion fail to concentrate on the real problem they have. This was brutally recalled by the senior parliamentarian at the workshop, who put the issue of the mismanagement of fisheries in the general context of food production in Senegal. He recalled that only a few years ago, his country allowed its own rice production to be destroyed by cheap imports from Taiwan, only to be hit a few years later with massive price increases, which have put the now-imported staple out of the reach of most of his compatriots. And he warned that the whales-eat-our-fish issue could have a similar effect by diverting attention from the task of putting Senegalese fisheries on a sustainable track.

NOT THE FISHERIES COMMITTEE[467]

~~~~~~~~~~~~~~~~~~~~~~~~~~~~

IT WAS ONLY A FILM

IN ONE OF the few funny scenes in the 1992 film *The Distinguished Gentleman*, the shady character played by Eddie Murphy, having tricked his way into the U.S. House of Representatives, is told that he is to be assigned to the "Fisheries Committee." His response, reflecting dismay at his instantly diminishing prospects for acquiring bribes, is a plaintive cry: "Not the Fisheries Committee, please!" And the American public laughs, because it is obvious that being assigned to a congressional fisheries committee is the worst thing that could happen in that country to any aspiring politician.

I don't think this line would be funny in Iceland or Norway, where fisheries matter. In the United States, they don't. In fact, the fisheries in the United States reportedly contribute as much to GDP as nail salons. Like other countries around the North Atlantic, however, the United States has a long history of fishing, starting with Native American fisheries,[468] but the fisheries became really important in the colonial period. In fact, in the 18th century, Boston became one of the bases from which Atlantic cod were eventually decimated.[469] At the same time, the northern coast of the Gulf of Mexico became the staging area from which shrimp trawling developed and took over the world.[470] However, it is through its Pacific fisheries that the United States, after World War II, had the largest influence on the fisheries of the world.[471]

TWO VERSIONS OF MAXIMUM SUSTAINABLE YIELD

THE CONCEPT THAT best represented the influence of the United States on international fisheries was maximum sustainable yield (MSY), invented by Wilbert M. Chapman. A biologist at the University of Washington, he became enormously influential as special assistant to the Secretary of State, the perch from which he ended up shaping the post–World War II foreign policy of the United States in the Pacific.

The dual challenge he faced was, on the one hand, to prevent the Latin American countries, whose waters harbored the tuna targeted by U.S. fisheries, from closing their waters to U.S. tuna vessels and, on the other hand, to simultaneously prevent Japan from accessing the salmon-rich waters off Alaska.

To justify this, Chapman drew a bell-shaped curve where the top was supposed to represent the MSY. To the left of this MSY was "underfishing," or the sin of letting tuna "die of old age" in the Southeast Pacific. To the right of the MSY was "overfishing," the sin that Japanese fishers were supposedly committing in Alaska, where they took salmon that U.S. fishers preferred to catch themselves.

Never mind that this curve had no biological foundation whatsoever and that it was never published in a scientific journal.[472] But the curve fulfilled its political role: the United States beat back attempts by Peru and other Latin American countries to extend their jurisdiction into large "Patriotic Seas" and expanded its tuna fishing in the southeastern Pacific, while at the same time Japan's fishing in Alaska was severely limited.

It was only a decade later that an approach, soundly based on theoretical ecology, was developed by Milner B. Schaefer to model the response of a single-species fish population to fishing—that is, to derive MSY scientifically and to provide criteria for defining over- and underfishing based on real data.[473]

However, Chapman's trick had worked, and the political version of MSY infested international relations. At the insistence of the United States, the MSY concept became part of the United Nations Convention on the Law of the Sea (UNCLOS)—which the United States then didn't sign.

Although to this day the United States has not ratified the UNCLOS, in the early 1980s it had to finally accept the declaration of EEZs by various countries, which affected the U.S. long-range tuna fisheries. At the same time, however, the UNCLOS enabled the United States to declare EEZs, meaning that it could ban Japan, eastern European countries, and other countries from fishing along the U.S. coasts.

By this time, U.S. coastal fish populations had been largely depleted by extensive foreign fishing. U.S. fisheries managers should have let these populations rebuild, but they did not; instead, they subsidized the expansion of their domestic coastal fleet,[474] which, in the 1990s, worsened the overfishing caused by the earlier, foreign fleets.[475]

Thus, the United States soon found itself in a situation similar to that of other industrialized countries—for example, Canada and the countries of the European Union. However, the institutional response in the United States was different: lawmakers passed legislation that effectively worked against overfishing. Specifically, in 1996, they modified the Magnuson–Stevens Fishery Conservation and Management Act of 1976, which had extended U.S. jurisdiction to 200 nautical miles and decentralized fisheries management by creating eight Regional Fishery Management Councils, from the North Pacific and New England in the north to the Gulf of Mexico and the Caribbean in the south.[476] The crucial change was that fisheries managers were mandated to rebuild overfished fish populations so that their biomass (B) would allow MSY to be taken (or to rebuild populations to B_{MSY}). In 2006, the proposed quotas (or permissible catches) were lowered so that the exploited populations would be rebuilt to B_{MSY} within ten years.

It is unfortunate that the 1980s, when many U.S. fish populations were already badly depleted, are often used as baselines for rebuilding. Still, the principle of the "reauthorized," 2006 version of the Magnuson–Stevens Act was bold in its implication that it is not legitimate (or, in fact, legal) to maintain a fish population so depressed that it cannot generate the maximum benefits for people.

Conservationists also generally go along with the intention of the Magnuson–Stevens Act, because rebuilding a fish population to the level that can generate MSY (the real one, not the fake one that W.M. Chapman invented) increases the biomass and diversity of fish in the water and makes marine ecosystems more resilient.

Currently, near the end of the 21st century's second decade, a large number of fish populations around the United States have regained reasonable levels of abundance. Because of both the Magnuson–Stevens Act and the prudent management practiced in most areas covered by the U.S. Regional Fishery Management Councils (except for New England, where overfishing continues to be rampant), the United States has relatively healthy fisheries.

Another major reason for the health of the fisheries is that U.S. fisheries management is rule-based (as prescribed by the Magnuson–Stevens Act) and not subject to what in Canada or the European Union would be called "ministerial discretion," or political interventions by high-ranking officials, which preclude science-based fisheries management. Such interventions are akin to a country's justice minister overruling a judge in an important trial, as opposed to working with the legislature on establishing laws (rules) that then apply to everyone and are applied and enforced by judges (or fisheries managers).

PICKING AT ONE'S SEAFOOD[477]

U.S. FISHERIES ALSO have a strong connection to "foodism." There is strong public interest in where food comes from, how it was harvested and by whom, and similar questions that were neglected before. This was like forging a new relationship with seafood, in which the delight of discovering a new lover is followed by the discovery of his or her issues, which eventually leads to a reassessment of the relationship, however painful this might be.

In the United States, the new relationship to seafood broadly expanded the choices that had been previously available. Seafood is tasty yet light and widely claimed to be good for one's heart and brain. But then, when it turned out there were health, legal, and ethical issues associated with seafood, it gradually lost its novelty.

The health issues are quickly summarized here: while the benefits of fish-derived omega-3s, often touted as good for the heart, are dubious[478] and in many cases not markedly superior to those of nuts, flax seeds, or leafy greens, the role of fish as a source of mercury, dioxin, and other pollutants is well established, and these can reach very high concentrations in some species, notably tuna[479] and farmed salmon.[480]

A friend might also tell you about the legal issue: many fish are caught illegally, or above quota, both of which can contribute to overfishing. Another friend might tell you that the fish you eat in restaurants are frequently misidentified, often deliberately.[481] In other words, you do not know where they are from or even what they are.

Then there are the ethical issues, some diffuse and easily ignored, others insistent and increasingly affecting our relationship with seafood. The fish sold in the United States and other developed countries with large markets, such as in Japan, Spain, or France, are mostly imported, since these countries have long overfished their own coastal resources. In fact, fish is a totally globalized commodity, consumed mostly on continents other than where they were caught.

And it is mainly from the waters of developing countries—many with rampant malnutrition—that most of the imports for the markets of developed countries originate. Because the world catch of wild fish is declining (it is![482]), the increased consumption of fish in rich countries means that less fish is available to supply local markets in poorer countries, thus deepening food insecurity and malnutrition.

No problem, we are told. We are going to farm the fish we need for "food security." But there is a hitch: salmon and many other farmed fish are carnivorous, and farming them involves feeding them with animal flesh, just as farming mountain lions would. For fish, the animal flesh, supplied in the form of pellets, consists of ground-up sardines, anchovies, mackerels, and other edible fish caught mainly—you guessed it—in developing countries. About three to four pounds of ground-up small fishes are required to produce one pound of farmed salmon. Thus, the more farmed fish we produce, the less fish there actually is.

However, this doesn't even begin to address the issue that the small fish we remove from the sea to feed farmed salmon, pigs, and chickens (fish-eating chickens!) are important prey of beloved animals like majestic seabirds and frolicking dolphins. In the Mediterranean, where tuna "farming" is now widely practiced, the removal of small fish for tuna fattening has been so extensive that the "common dolphin" is becoming rare, with individual dolphins showing protruding ribcages like the mangy dogs one can see at the edge of dusty settlements in some developing countries.

These are real issues, capable of inducing concern and guilt in at least 5 percent of the population. To assuage this guilt, various ritual objects and behaviors have been devised, of which wallet cards are the most noticeable. These are small, folding cards illustrated with those fish that are "good to eat" (green), those that are "to be eaten with caution"—with lips pursed?—(yellow), and those that are "not to be eaten" (red). The corresponding emotions are feeling smug (green), slightly risqué (yellow), or really naughty (red). However, whether these cards and similar consumer-oriented initiatives ever achieve their stated goal of decreasing overfishing has rarely, if ever, been evaluated.[483]

There is an alternative to these guilt-based approaches, and it involves another emotion: shame. Corporations, which cannot feel guilt,[484] can be publicly shamed and forced to change their behavior when the public pressure becomes too strong.[485]

AN UNDISTINGUISHED GENTLEMAN

SHAME, OR RATHER shamelessness, brings us back to where we began, when a trickster managed to insinuate himself into the U.S. government. That trickster was, fortunately, an improbable film character. However, in November 2016, a very undistinguished non-gentleman was elected to the U.S. presidency, and while *The Distinguished Gentleman* may have been funny, this election was not.

Notably, the non-gentleman who poses as U.S. president made it his mission to demolish all the environmental protections that he can and to foster short-termism ("clean coal" anyone?) over sustainability. Thus, the marine reserve around the northwestern Hawaiian Islands—declared by former President G.W. Bush and expanded by his successor, former President B.H. Obama—that initiated a global movement toward the creation of marine reserves[486] is, at the time of writing (September 2017), in danger of being re-opened to fisheries.[487]

We can only hope that the near-invisibility of fisheries in the U.S. political scene will shield U.S. fisheries regulations from the demolition crew and that the "fisheries committee" continues to do its good work.

MY PERSONAL ODYSSEY I: ON BECOMING A CANADIAN FISHERIES SCIENTIST[488]

LANDING IN HOT SOUP

ALTHOUGH I FIRST began as a professor of fisheries at the University of British Columbia (UBC) in September 1994, I will start this essay in the fall of 1995, when I officially "landed" in Canada. I had already noted that people really are more polite in Canada—for example, on buses—than in most other countries that I had then visited, but I was still discovering its culture, or at least that part that one could see and hear in Vancouver.

One evening in the fall of 1995, I was invited to one of a multitude of events that, on the West Coast, tend to be devoted to the salmon fisheries. That event—an evening lecture—was attended by John Fraser, a former federal minister of fisheries. Near the end of the event, someone proposed that the assembled, mostly fisheries types and other academics, should raise our glasses and give a toast to the current federal minister of fisheries, Brian Tobin, "for his courageous stand against the rapacious Spaniards who were taking all of our fish." Being a teetotaler, I did not have a proper glass to raise, but as it turned out, I did not need one: somebody

(I think, without being sure, that it was Mr. Fraser himself) stood up and said that he did not agree: Mr. Tobin had not represented Canadian values when he ordered the Canadian Navy to confront a Spanish vessel in international waters, blaming the Spaniards for all the ills of Canadian fisheries and stirring up xenophobic emotions. Mr. Tobin was not celebrated that night, and I learned that not all Canadians approved of the propaganda campaign then in full swing in the Canadian media.[489]

SALMON FARMING AND ITS PARASITES

IN THE 1990S, British Columbia was beginning its experiment with farming Atlantic salmon, which became a very divisive issue in the province. Having worked on aquaculture since the early 1970s, I was fairly knowledgeable about its different forms throughout the world[490] and was puzzled by the Canadian Department of Fisheries and Oceans' (DFO) denials of the evident problems with regular, massive escapes of farmed Atlantic salmon and their transmission of parasites to wild Pacific salmon.

Most of the controversies surrounding this issue swirled around Alexandra Morton, an independent researcher who had raised the alarm about the impact of salmon farms on the surrounding ecosystem in the Broughton Archipelago, where she lived and studied killer whales.[491]

Although my research is usually performed in offices (I like to re-analyze other people's data[492]), I knew that, in this case, it was better to see for myself. Thus, with my wife, Sandra, a colleague whose PhD was on fish parasites, I visited Alexandra in July 2004, taking a flight from Vancouver to Port Hardy on Vancouver Island, then a ferry to the Broughton Archipelago and a water taxi to Echo Bay, where Alexandra lived in a sprawling old house with a view of the salmon farms on the other side of the bay.

Alexandra immediately took us for a ride on her tiny motorboat, which dragged a net in the shallows. We caught about a hundred little fish, mainly young salmon. Each and every one of them carried large sea lice on their bodies, of a size corresponding to a dinner plate on the chest of a human; some had two lice attached to them, and there was little doubt that these lice had their origin in the salmon farms across Echo Bay. Later in the evening, Alexandra showed us the scientific publications she used to identify and stage the lice, becoming in the process the foremost expert in British Columbia on these infestations.[493]

There is nothing like evidence such as this to fortify one's stand on an issue. Later, when visiting the DFO research vessel *Ricker*, I heard no less than the director of research of the DFO's Pacific Research Station in Nanaimo say publicly, before senior members of her staff, that Alexandra had manually "spiked" young salmon with individual parasites. I knew that this was a horrible lie and just the beginning of a concerted effort to discredit Alexandra's findings.

And when I heard that the DFO had conducted a cruise in the Broughton on R/V *Ricker* and found "no parasitized young salmon," I knew that when you do not want to find young salmon with sea lice, you go sampling with a research vessel of 190 feet that cannot operate in shallow waters, the only place where young salmon can be found.[494]

I fear that the DFO did not cover itself with glory during this period. And still, the sad saga of denying evidence that was staring everybody in the face continued, as exemplified by the treatment of their top geneticist, Dr. Kristi Miller. Dr. Miller documented the presence of an imported disease-causing virus in farmed salmon[495] but was not permitted to talk to the Cohen Commission, charged by the Canadian government to investigate the reasons for the decline of wild salmon in British Columbia.[496]

FISHERIES CATCHES ALONG THE BC COAST

I ONCE EXPERIENCED for myself the DFO's *"omertà,"* when I asked for official catch data on the BC fisheries and was rebuffed (I was at that time the director of UBC's Fisheries Centre) by one of its leading scientists, who wrote that DFO would not provide the data because I would deliberately misinterpret them.[497]

Yet, at the time, all I was doing was trying to improve and update the catch data in the PhD thesis of one of my first graduate students, Scott Wallace, who had reconstructed marine catches in BC from 1873 to 2011.[498] The job of improving, updating, and publicizing BC catch data was eventually done by Cameron Ainsworth, another UBC Fisheries Centre graduate[499] who, building on Scott's work and other sources, demonstrated that a large fraction of fisheries catches in BC, notably recreational catches and the catch by First Nations, do not make it into DFO statistics[500] and are thus not included in the catch statistics that Canada reports annually to the Food and Agricultural Organization of the United Nations (FAO).[501]

Thankfully, at the time of writing, it seems that the present federal government will follow up on the policy of transparency and openness that it has committed to.[502] I hope this policy also reaches the BC branches of DFO, which have, to date, been anything but transparent or open. However, this will occur only if there is a continuous push and pressure from the nongovernmental organization community—for example, from the David Suzuki Foundation (DSF), which does very good work in BC. I might be biased in choosing this example, however, as two of my former graduate students have done small consultancies for the DSF,[503] and Sarika Cullis-Suzuki, David's youngest daughter, completed her MSc studies under my supervision.[504] This brings us to the Canadian Arctic, about which Sarika and her father have made an excellent film.[505]

FREEZING OUT THE CANADIAN ARCTIC

CANADA BOASTS OF having shores on three oceans, the Atlantic, the Pacific, and the Arctic, but not much is said or written about its Arctic fisheries, which is strange given that the overwhelming majority of Canadian Inuit communities are coastal.

It is not that Inuit relationships to the sea in general and to fishing in particular have not been studied. On the contrary, the DFO office in Winnipeg, responsible for the Arctic, has piles of reports on marine fisheries and fish consumption by Inuit communities along the Arctic coast, from Hudson Bay in Nunavut to the Beaufort Sea in the Northwest Territories. The problem is only that these reports were never used to construct a comprehensive time series of fisheries catches from the Canadian Arctic and report on these fisheries to the FAO. Thus, in the mid-2000s, I encouraged Shawn Booth, then a research assistant with the Sea Around Us, to team up with a colleague, Paul Watts, who knew about the "hidden" DFO reports, to weave the marine fish consumption data from various Inuit settlements over the years into a coherent time series for the whole of Arctic Canada.

The idea was that, absent imports from the south of Canada, these consumption figures can replace catch statistics. The preliminary report that came out of this work in 2007[506] later became part of a comprehensive article published in *Polar Biology*—a serious scientific journal—that in addition to the Canadian Arctic covered the fisheries of Arctic Alaska and the Siberian coast from the Chukchi Sea in the east to the Kara Sea in the west.[507]

In that article, we pointed out that like Canada, Russia and the United States do not report the fisheries catches of their "people of the North" (the term used in Russia) to the FAO, thus giving the impression (at least in the fisheries world) that these peoples do not exist. But they do, and in Canada they collectively have caught over 2.2 million pounds of fish per year since mid-1975, mainly char (*Salvelinus alpinus*), a salmonlike species.

The reception of that article was interesting: most journalists got our point that in Canada an entire sector of the Arctic economy should not be ignored by the same government that proclaimed—as the Harper government did—that Canadians must assert their sovereignty over the Arctic.[508]

Then there was a phone call from a bureaucrat in the Nunavut Department of the Environment asking me to send her the paper in question. I send copies of papers to anyone who asks for them, but I wondered why she didn't look at *Polar Biology*. I was not ready for the response: "We don't have this journal—we can't afford it."

Because the editor-in-chief of that journal, Professor Gotthilf Hempel, had been my doctoral thesis supervisor, I was able to arrange for the government of Nunavut to get a free subscription to the journal. But you have to think about it: the government of a Canadian territory with an area larger than Mexico cannot afford a scientific journal that deals precisely with the issues it must address (notably natural resources management and climate change).

I discovered something even more troubling from the "catch reconstruction" for Arctic Canada. In the 1950s and 1960s, the consumption (and hence catches) of char was two to three times higher than in later decades. The additional catch was for the Inuit's sled dogs, which were fed frozen char, apparently an excellent dog fuel.

Then I learned, to my horror, that from 1960 to 1970 the Royal Canadian Mounted Police had killed the dogs to force the Inuit to abandon their peregrinations and live in "proper" settlements, which a benevolent government would then provide with all the amenities of "proper" Canadian life.[509] We now know that these amenities did not necessarily include running water (except from the leaky roofs). But the first part of the program—the killing of the dogs—was completed. The Far North now has the highest rate of teenage drug addiction and suicide in Canada. I wonder if killing all those dogs did not contribute, at least in part, to this state—which brings us to a major issue in Canadian history…

DEALING WITH CANADA'S HISTORY

INTEGRATING INTO CANADA, as I did from the mid-1990s onward, implies dealing with its history. For me, this involved reflecting not only on the history of its fisheries[510] but also on the history of its Inuit and First Nations and the horrors that befell their children in Canada's system of residential schools. This is the reason I accepted an improbable invitation to join Canada's Truth and Reconciliation Commission (2008–2015) as an Honorary Witness.[511] This also allowed me to participate in several events, in Vancouver and elsewhere in Canada, at which older victims of the residential school system presented searing testimonies of what they had endured as children and of what their own children endured as a result. The Reconciliation Pole recently erected on the UBC campus expresses all of this,[512] as well as the hope that, as a society, we will address this lingering pain.

ECOSYSTEMS, MODELS, AND SALMON TREES

WHEN I CAME to UBC, I had something to contribute to my scientific colleagues in British Columbia: I had co-developed software for ecosystem modeling called Ecopath, which, until then, had been applied mainly in the tropics.[513] I taught the use of this software and the underlying principles in the first year I was at UBC. Then, in 1995, I convened a week-long workshop, held November 6–10, to apply Ecopath to marine ecosystems of the Northeast Pacific, notably the Strait of Georgia off the BC coast. The Strait of Georgia had been studied extensively, but the data from different studies, covering the entire food web from plankton to marine mammals with all the fish and invertebrates in between, had never been studied.

At the workshop, held jointly with my colleague Villy Christensen, we asked scientists who had worked their entire career on "their" group of animals (orcas, seals, salmon, crabs, or zooplankton) to express their estimates of abundances and food consumption using the same currency so that they could be linked with each other—as they are in the sea. It worked: the first Ecopath models of the Northeast Pacific ecosystem were created during that week.[514] That same week, Carl Walters, who had initially remained distant from this modeling exercise, showed how Ecopath, which, while comprehensive, was only a static representation of an ecosystem, could be made dynamic and represent change, for example, as the result of more fishing or a natural decrease in plankton production.[515] This was a real breakthrough in ecosystem modeling.

That magic week in late 1995 became the basis for what may be seen as the best ten years of the UBC Fisheries Centre, which became a modeling powerhouse. Colleagues came from overseas to learn how to use what then became known as Ecopath with Ecosim (EwE), now the standard software used by fisheries scientists all over the world to simulate the dynamics of the ecosystems in which fisheries are embedded.[516]

At that time, the Fisheries Centre also had an excellent relationship with the now-defunct Aboriginal Fisheries Commission and other First Nations groups, which enabled me to supervise a First Nations student named Steven Watkinson. His master's thesis used a spatial version of

Ecopath to study the spread of salmon-borne, marine-derived nitrogen into the forests of the Central Coast of BC, mainly via predation by bears,[517] leading to the evocative concept of "salmon trees."[518] Steven also participated in an attempt to retrieve as many fish names as we could[519] in two First Nation languages (Haida and Tsimshian) and to disseminate them via FishBase, the global online encyclopedia of fishes.[520]

BEING CANADIAN ENOUGH?

UNFORTUNATELY, THE COLLABORATION and comity that had led to the blossoming of UBC's Fisheries Centre did not last. Some colleagues criticized my continued commitment to also working on tropical/developing countries through the Sea Around Us, a research project that I started in mid-1999 with funding from a large U.S. philanthropic foundation. Never mind that I had been hired by UBC to work with international students and to assist Canadian fisheries development projects abroad, that I led many Canadian graduate students to MSc or PhD degrees,[521] and that the Sea Around Us employed dozens of Canadians.

There were colleagues who suggested that I should not devote so much of my time to non-Canadian issues, given that I was working in a provincial university. These colleagues had not understood that science is international and that the prominence and growing reputation of the Fisheries Centre was in large part due to a methodology developed elsewhere, for Ecopath had first emerged in Hawaii,[522] was further developed in the Philippines,[523] and was widely applied in Mexico[524] before it arrived on Canadian shores.

This situation and various other issues led to a crisis that the UBC dean of science thought he could resolve with a sledgehammer. Thus, the Fisheries Centre is no more.

Returning to the title of this essay: I became a real Canadian fisheries scientist only in 2016. I had become a Canadian citizen quite late, in 2014,[525] when a judge decided that I was OK even though I had never been able to accumulate the required 1,460 days (four years within a six-year period of residence in Canada) because I travel too much.[526] Then, in September 2016, I was named one of the "Legends of Canadian Fisheries Science and Management" by the kind members of the Canadian Chapter of the American Fisheries Society. As the previous nominees include such giants as David Schindler and Bill Ricker, things are not going to get any better. So I stop here.

MY PERSONAL ODYSSEY II: TOWARD A CONSERVATION ETHIC FOR THE SEAS[527]

A WEIRD BACKGROUND FOR A MARINE SCIENTIST?

ALTHOUGH I WAS born in Paris, I grew up in La Chaux-de-Fonds, a little town in the French-speaking part of Switzerland, in the Jura Mountains, where cows roam freely but not very far, because they have bells. I did not have a "normal" youth,[528] but I did have hamsters, a goldfish, and sometimes a dog. However, I did not have the intimate connections with Nature that some well-known biologists enjoyed. I was into books and ideas, never a naturalist. I tend to see patterns in data, but not in raw Nature.

At sixteen, I dropped out of high school and went to work in Germany for a year as a "diaconic helper" for six months in an asylum run by the Lutheran Church and another six months in a city hospital, which cured me forever of the religious delusions common in juveniles. Instead, I realized that I needed to go back to school, and this I did: for four years, I attended evening school from 5 to 9 p.m., five times a week, while working low-level jobs in a paint factory, a brush factory, and other factories during the day to support myself.

Nature, never in the foreground, receded even farther into the background of my life.

Then, in the spring of 1969, I graduated and went to the United States to connect with my father and his family for the first time, much as I had reconnected with my mother and her family three years before in Paris. As the son of a Frenchwoman and an African-American G.I., I had previously been aware that I was biracial (and there was always somebody to remind me, lest I forgot), but I was not ready to be part of a group. In the United States, I became assimilated into one ("African-Americans") that was still engaged in the fight for civil rights and its various ramifications. I came out of this experience more confused than ever but convinced that I should somehow join in the struggle of people of color. I decided I would not live in Europe after my studies.

Thus, when I began my studies at Kiel University, I aimed to learn something "useful," something that would enable me to work in developing countries. I obtained permission to do a double major in biology and agronomy, but unreformed, old Nazi professors (real, not metaphoric ones!) drove me out of agronomy. They also denigrated and lied about Rachel Carson and her book *Silent Spring*, which I had just read and which greatly impressed me.

Marine science offered an alternative mix of useful and neat science, and Kiel University was a good place for it: you could learn classical fisheries science and marine biology at the very place where, in the late 19th century, Victor Hansen and Karl Möbius founded planktology and (benthic) community ecology, respectively. The courses available were a nice combination of theory and practice—that is, laboratory and at-sea work, the latter being my first real introduction to the marine realm. The fisheries work was conceptually easy: we were to study and understand the resource species so that they could be "managed." There were "managers" somewhere who would use our results to get the most out of the resources in continuous fashion.

Here are two things that impressed me during my studies: First, my six months (in 1971) in Ghana to study a coastal lagoon, near the port city of Tema. I learned all about the little lagoon, which supported an artisanal fishery for blackchin tilapia,[529] and even discovered a new species of parasite in their mouths.[530] Now, almost fifty years later, the lagoon is inside Tema, and the tilapia, size-wise, have turned into guppies. It was also then that I got my first sunburn and learned that I was European, not African. Second, my six weeks (in 1973) onboard a giant factory ship turned research vessel, surveying cod off Newfoundland and Labrador. These were the heydays of the cod fishery (which collapsed less than two decades later). We were fishing at 1,600-foot depths, with trawls capable of lifting a boulder the size of a Volkswagen. Now I understand: we did not know what we were doing.

In 1974, I obtained a "Diplom," the German equivalent of a master's degree, and I was hired by the German international development agency (GTZ) to work in Tanzania. I learned about Indian Ocean fish in a Frankfurt museum and sat through four months of Swahili, enough to hold a simple conversation. Then, in mid-1975, I was shipped to Indonesia to help introduce trawling to the country.

In Indonesia, I did the standard work of foreign fisheries "experts" working in developing countries, that is, helping to "develop" fisheries. This consisted mainly of conducting surveys to estimate the then-largely unexploited demersal fish biomass of western Indonesia and writing reports about how much was there and how much could be taken. I later realized that the people acting on this information were the staff of development banks, who used our estimates of "potential yields" (i.e., "possibly maximum sustainable yield, but be careful") to justify loans to countries with which vast trawler fleets were purchased.

The Manila-based Asian Development Bank and other international development agencies used our potential yield estimates (which, with hindsight, were far too high) to make decisions about loans aimed at developing industrial fisheries but did not much care about their sustainability (the word wasn't used then) or about the impact of fishing on ecosystems ("ecosystems" did not exist then for fisheries science). On the ground, however, we were transforming the coastal ecosystem from a soft coral system to a muddy bottom system. Of the many scientific challenges at the time, three now stick out: (1) we were "terraforming" the sea (but did not know it), (2) we were ignoring small-scale fishers (but did not care), and (3) we were dealing with more species than we could handle.

I chose the last of these problems as my research challenge. The two years in Indonesia passed quickly, and I then returned to Germany with my head full of ideas for how to improve fish stock assessments in the tropics. I earned a PhD working out some of these concepts and also teamed up with colleagues who helped me program some of the more outlandish ideas—for example, a program called ELEFAN (Electronic LEngth Frequency ANalysis) that was used for the estimation of growth from length–frequency data on fish and that ran on microcomputers such as the Apple II, then fresh from Steve Jobs and Steve Wozniak's garage.

In 1979, I was back in Southeast Asia, this time in the Philippines, at the International Center for Living Aquatic Resource Management (ICLARM). This institution, founded two years earlier by the Rockefeller Foundation, was to do for the ocean what the green revolution had done for the land (increasing yields, the panacea in those days). For me, this meant teaching my newly developed methods and concepts as tools of "empowerment" throughout the tropical world. Thus, I got to know hundreds of colleagues on five continents and found out that they all had similar concerns.

Traveling as I did and crossing cultures and languages helped me see similarities where others saw differences. In the 1980s, the artisanal fisheries of Senegal, in northwestern Africa, were booming and were a source of wealth, and those of the Philippines were already in deep trouble. However, unlike many anthropologists, I understood that this difference was not due to differences in the countries' social organization, or "Asianness" versus "Africanness," but to contingencies of development, such as when development began. Now, Senegalese fisheries are in the same trouble as those in the Philippines. This required a theoretical explanation, which I endeavored to develop, and which brings us to our first digression.

DIGRESSION: A SCIENTIFIC REVOLUTION IN THE 1980s AND 1990s

JUST AS THE marine sciences were transformed in the 1960s by the theory of plate tectonics, which provided a powerful explanation for a number of geological and biogeographical phenomena, anthropology was revolutionized by the realization, in the 1980s and 1990s, that we are a modern species, only 150,000 to 200,000 years old, not derived from the various hominid species outside of Africa (e.g., the Neanderthals).

Moreover, we now have good evidence that all non-Africans alive today are the descendants of a single small (<1,000) group of humans that left Africa for Asia by crossing into what is now Yemen about 70,000 years ago. These migrants had all the mental equipment that enabled them to gradually people the whole world and later overwhelm it. Thus, all humans are closely related and have only recently separated, so that even the identification of features of an ancestral common language can be envisaged,[531] not to mention the many initiatives documenting our genetic affinities.

The take-home message is simply that there is a good scientific basis for assuming that we are similar and react to similar challenges in a similar fashion. This is obviously the basis for evolutionary psychology[532] and other initiatives to put the various social sciences on a solid Darwinian basis.

Now, what is it that people do? Or, what is our niche? Bill Rees, of the University of British Columbia, and others define our species as "patch disturbers":[533] we mess up a place and then move on. Elephants do this as well, and where there are too many of them, they will reduce the vegetation covering and overall biodiversity in a place, just as we do.

But we are not elephants. For a long time, our pre-human ancestors were simply one of the many prey on which the great cats and other predators relied; we were then truly, as the otherwise meaningless phrase goes, "part of the ecosystem." But we had those big brains, and we planned and did things in groups and gradually ceased to be the prey and became serious competitors for the great carnivores.

As we hugged coastlines, our main initial routes of expansion, we gradually peopled the world, even the Americas, through the kelp highway, as new scholarship suggests,[534] and finally built the sophisticated boats that enabled the peopling of Polynesia.

We wiped out all the big herbivores we could access: the large marsupials in Australia 50,000 years ago, immediately upon our arrival there; mammoths in Eurasia about 20,000 years ago; forty or so species of large mammals in North America about 11,000 years ago;[535] and the eleven species of moas in New Zealand, also right after the first Polynesians arrived. There is a huge amount of debate about this, mainly because the notion of ancient peoples having such destructive power doesn't sit well with everyone.

Then we began what I think is definitely our move out of the (natural) ecosystem: about 10,000 years ago, we began farming, and the hunter-gatherers throughout the world began their long decline into marginalization. No more patch disturbing? Not so: rather than going after the large animals that lived off the land, we went after the land itself.

I'm referring here to David Montgomery's excellent book *Dirt: The Erosion of Civilizations*,[536] which demonstrates that every succeeding civilization derived its power from the fertility—hence the soils—of its core region. Whether we are looking at the Babylonians or the Hittites, their hegemony lasted only a few hundred years—as long as it took for their soils to be ruined.

The apparent exception to this sad story is ancient Egypt, which lasted millennia, but whose soil was replaced every year by the Nile and, of course, had to be gotten rid of. Now, with the Aswan Dam, we can expect Egypt to go the way of Babylon.

BACK TO THE 1990s

IN THE 1990S, there was an administrative crisis at ICLARM (the longevity of research institutions may also be inherently limited), and I was offered a position at UBC. But then, I asked myself, should I go on researching and teaching fish population dynamics and ELEFAN and other nifty little tricks? No: my kind of simple-minded work was too unsophisticated for Canada (where we make sophisticated messes), and I had learned from the cod collapse that Canadians don't know what they are doing either, and the political system, if not prodded, won't do anything. I found this situation was similar, after all, to the situation in developing countries, where the colleges where I taught had no influence on decision-making in fisheries, making development-*cum*-management a parody.

But in the Philippines, I had also lived through the people-power revolution that overthrew the conjugal dictatorship of Ferdinand and Imelda ("she-of-the-shoes") Marcos in 1986, confirming something I also saw in May 1968 in Paris: people can force government and its agencies to do what they don't want to do (because they have been captured by industry lobbyists).

So, at one of the first fisheries meetings I went to in my new position, I offered my services to a group of baffled representatives of environmental nongovernmental organizations huddling among themselves instead of working the crowds.[537] This led me to participate with the World Wide Fund for Nature (WWF) in the work leading to the creation of the Marine Stewardship Council (MSC) and to begin a series of public lectures given in all corners of Canada, the United States, and other countries and continents.

This is also the time when I was lucky enough to publish a series of papers that had a certain success, notably on shifting baselines (see the essay titled "The Shifting Baseline Syndrome of Fisheries") and on fishing down marine food webs.[538] In the fall of 1997, the Pew Charitable Trusts invited me to join a one-day workshop in Philadelphia with a small group of marine scientists to discuss how to assess the health of the ocean. The other scientists all said that "ocean health" was not a scientific concept and that they would need more data (this is the kind of thing that causes people to think scientists are useless). I proposed instead that the world marine fisheries statistics from the Food and Agriculture Organization of the United Nations (FAO), gathered from all oceans for almost five decades, could be seen as a huge sampling program and that we should first analyze it. I carried the day.

Thus, at UBC, I began to assemble a brilliant group of people (Villy Christensen, Rashid Sumaila, Reg Watson, Deng Palomares, Dirk Zeller, and others) and to design a project called the Sea Around Us, named after Rachel Carson's best-selling book, to document the state of the oceans and to work with civil society to help slow down or reverse negative trends.

One of our first visible results was an analysis of world catch trends, which demonstrated that the world catch was declining. Although this decline could be inferred from negative stories everywhere, it had been masked for years by catch over-reporting by China.[539] This result occurred around the same time that Jeremy Jackson and his colleagues were showing that overfishing had been the *modus operandi* for millennia.[540] The papers of the late Ransom Myers were also hitting the news at this time.[541]

Jointly, these papers changed the view that fisheries are isolated affairs, failing separately from each other. Rather, we now realize that our entire mode of interaction with the sea is wrong, just as we don't believe anymore that this or that bank has failed but that the whole financial system has failed us.

CURRENT TRENDS

OVER THE YEARS, as we have made more maps showing fisheries as a global system, we have come to realize that the old patch-disturber has now almost completely disturbed the biggest patch on Earth, the oceans, through three expansions: (1) geographic, from the north to the south, (2) bathymetric, from shallow to deep waters, and (3) taxonomic, into new species we didn't eat before (see the essay titled "Duplicity and Ignorance in Fisheries").

We also came to realize that we almost never had real sustainability, defined as a given group of humans maintaining (or even coexisting with) an animal population, despite their ability to deplete it. We had even less of that oxymoronic concept "sustainable development."

The world of fisheries has been warped by the demand emanating from a few markets (the European Union, the United States, Japan, China), all with devastated local fish populations and insatiable appetites. In Europe and North

America, not only do we want to eat all the fish, but we also pick and choose sustainable fish from magic wallet cards so that we can feel good about ourselves. The MSC, in a move similar to Anakin Skywalker's, is certifying reduction fisheries, which turn catch into fishmeal or fish oil; in other words, it is certifying as "sustainable" the grinding up of perfectly edible fish to give to pigs to eat.[542] The result is that an ever-increasing tonnage of farmed salmon can be certified as "sustainable" because they are fed certified anchovies. Here is the old patch-disturber again, this time wrecking even the very language with which we think and express ourselves.

Clearly, we now consider the entire planet a patch that we can disturb before moving on, and many science fiction authors suggest that we will do just that if we can. Lester Brown has a beautiful article in *Scientific American* outlining a way to counter this with what he calls "Plan B."[543]

THE NEED TO UPDATE OUR OPERATING SYSTEM

TO CONCLUDE, NEVERTHELESS, on a positive note, I will briefly elaborate on the mental underpinnings of my version of Plan B, using the computer as an analog to our brain (the hardware).

Essentially, through our (pre-)history, we ran three successive operating systems on our computers/brains:

OS1: WE-ARE-PART-OF-THE-WORLD-AROUND-US

THIS SYSTEM IS embodied in various animist religions, based on the belief that we share with animals, plants, and inanimate objects an essence or spirit. This system emphasizes natural cycles and linkages between things, animals, and people, mediated by magic. A working system, but easily beaten by its successor.

OS2: WE-ARE-THE-MASTER-OF-THE-WORLD

THIS SYSTEM IS embodied in the revealed religions, in which a top boss (the deity) put us in charge on Earth, which we leave when we die. We do not share any features with animals (whose sole reason for existing is to be exploited by us), but with the deity, and we use ritual and magic to communicate with it. A delusional yet very robust system.

OS3: WE-KNOW-WHY-WE-ARE-PART-OF-THE-WORLD-AROUND-US

THIS SYSTEM IS the beginning of a scientific (or rational) worldview, deeply connected with the ideals of the Enlightenment (equality, human rights, and all that), but very fragile and easy to crash (see, for example, the Nazis, right-wingers, or religious fanatics of various kinds).

The key feature of OS3 is that it works on a set of basic beliefs that are not challenged in theory—for example, that all people on Earth are equal and should get a fair deal. This is not true in practice but is a shared belief that we use to judge practical events. This is also the foundation myth of modern democracies. Part of this foundation myth is that we should know things and that knowledge is a good thing, the basis for the peaceful coexistence of democracy and science.

Clearly, we will have to graft onto the fabric of democracy and of the Enlightenment itself a new set of values (also called rights) pertaining to our continued existence on this planet. Let's call this OS3 v2.0. This will involve a right to a fair share of the planet's resources; today this right is being negated, mainly by corporations, and by the inhabitants of a few countries eating up all the Earth's resources. This system will also include a right to learn about the world, rather than being brainwashed into belief systems that close minds.

Foremost about v2.0 is that it will always require any new venture to assess whether it is truly sustainable—that is, independent of fossil fuels and consumption of soils or other nonrenewable resources.

There is no reason why such an OS shouldn't work. Thus, I am not wholly pessimistic about the future: we have an operating system that, with an upgrade, could replace the one now failing us. I just hope that the upgrade occurs before it is too late, before we have completely trashed our planet.

MY PERSONAL ODYSSEY III: HAVING TO SCIENCE THE HELL OUT OF IT[544]

LET'S START IN THE EARLY 1970s

THE 1970S WERE, for fisheries science and scientists, a period of transition. The world's marine fisheries catch, which had been increasing in the 1950s and 1960s—to the degree that catch *per capita* increased despite massive population growth—transitioned to slower increases, punctuated by the spectacular collapse of the Peruvian anchoveta (*Engraulis ringens*).[545] This general trend led to a frantic search for new fishing grounds, with industrial trawl and tuna fisheries expanding, for example, onto the previously underexploited Patagonian and Sunda shelves and the Indian Ocean, respectively.[546]

In previous decades, most of what was known of the fisheries in what a report by the International Council for the Exploration of the Sea (ICES) called "other areas,"[547] many of them former colonies of European countries, was the work of taxonomists describing the strange fish that were caught, or anthropologists describing the strange traditions of small-scale fishers.[548,549] In the 1970s, however, the international aid programs and international banks that fueled this expansion of industrial fisheries needed some sort of assessments to justify their ventures (e.g., the Asian Development Bank, which "[c]ommencing with its first fisheries loans in 1969, ... has made loans of US$1,055 million to a total of 51 fisheries projects in 17 [developing countries]."[550] Fisheries science responded, led by the indefatigable John Gulland, a senior scientist at the Food and Agriculture Organization of the United Nations (FAO), and this is where my story begins.

Even before I completed my "Diplom" in Fisheries Science, Zoology, and Oceanography at the University of Kiel, Germany, I had decided to work in a tropical developing country, both because it was (and still is) not easy to be biracial in Europe, and because, as outlined in the previous essay, I wished to somehow "help." Following an intervention by my thesis adviser, Professor Gotthilf Hempel, I was hired by the German international development agency (GTZ) to work in Tanzania (this would not have occurred without that intervention, as I did not really fit the profile of a German fisheries expert). By way of training, GTZ sent me to learn about Indo-Pacific fish in Frankfurt's Senckenberg Museum and arranged for a Zanzibari émigré to teach me Swahili. I became able to speak Swahili but promptly forgot it because, in spring of 1975, GTZ sent me to Jakarta, Indonesia, the German-Tanzanian project I was initially hired for having failed to materialize.

INDONESIA: 1975 TO 1976

INDONESIA—THAT IS, THE Java Sea and the southern tip of the South China Sea—where we were tasked with introducing Indonesians to the joys of industrial-scale bottom trawling, is also where I first understood the scientific challenges that faced fisheries science in "other areas," where a multitude of species are being exploited by a multitude of fishers, using a multitude of gears. Throughout my time in Indonesia, I remained largely oblivious to the habitat damage that bottom trawling causes (at the time there was no conceptual language for that). However, I understood, as soon as the first trawl haul appeared on the deck of our research trawler, that for the species in the wiggling, variegated heap of fish and invertebrates on our deck, we could never obtain the detailed biological data that underpin stock assessments in northern temperate developed countries. Thus, the most abundant demersal species in our trawl survey off West Kalimantan (Borneo), in the southern South China Sea, the red filament threadfin bream (*Nemipterus marginatus*), made up only 1.2 percent of the catch. Clearly, this made yield-per-recruit analysis, the state of the art in population dynamics at the time, essentially useless.[551]

I realized then and there that for fisheries science to be able to deal with tropical biodiversity, it would require more than the application of the tricks and approximations that were fashionable in fisheries science at the time. Rather, it was a genuine scientific problem, and this explains the title of this article, adapted from the utterings of a fictional astronaut left behind on Mars, who realized that to survive until he could be rescued, he was "gonna have to science the sh*t out of it" (from the film *The Martian* based on the book of the same title).[552]

MY FIRST RESEARCH PROGRAM

IN JANUARY 1977, back at the Institute für Meereskunde at Kiel University, I initiated a research program (and doctorate thesis, again under Professor G. Hempel) consisting of:

1. Developing a computer-based method for estimating growth and derived parameters (total mortality and gear selection) from length-frequency data (which are easy to collect), to replace the subjective pencil-and-paper methods then in use.
2. Understanding the basic drivers of the growth of fish and aquatic invertebrates so that the growth parameters of under-studied species could be inferred.
3. Assembling all published estimates of natural mortality (M) in fish and using them jointly with easy-to-estimate correlates to predict M in any species, anywhere.

Item 3, essentially a meta-analysis done at a time when I did not know the term, was the easiest of these three tasks. In 1978, at an ICES meeting in Copenhagen, I anxiously presented a first version of an empirical equation based on 122 estimates of M in a wide range of cold- and warm-water fish species that linked M to their growth parameters (asymptotic length and growth coefficient of the von Bertalanffy growth equation) and the mean water temperature of their habitat.[553] The audience included several of the reigning princes of fisheries science at the time—for example, Rodney Jones, David Cushing, and, I think, even Ray Beverton, our king. I survived my presentation, as all questions were constructive and none hostile. Later, the analysis was expanded to 175 cases and published, with editorial help from Keith Brander, in the venerable *Journal du Conseil pour l'Exploration de la Mer*.[554] This was to become my first heavily cited paper. It also got the attention of John Gulland, whom I had met previously at a workshop in Thailand. John then tried to get me to work for the FAO; I was willing, but we failed, because I had a French passport, but no support from France. For a while, he became a mentor nevertheless.

Item 1 was a bit more difficult; I had sought the help of a colleague skilled in mathematics, statistics, and programming (areas where I was never more than mediocre, notwithstanding having authored a textbook full of equations),[555] and predictably, he developed the outlines of a complex software that decomposed length-frequency (LF) distributions, assumed to be normal, assigned relative ages to them, and then fitted the growth curve, which minimized the sum of least squared deviations. Approaches of this type succeeded later,[556,557] with one, called MULTIFAN, finding some practical use,[558,559] though apparently mainly for tuna.[560] However, these approaches seemed far too complex at that time for the user group that I had in mind. Moreover, they all assumed a previous knowledge of whether the LF samples at hand represented one or two cohorts per year (and their relative strengths, as inferred—for example, from catch per effort data), which is precisely what was not known about tropical fish populations.

Therefore, right after I began working at the International Center for Living Aquatic Resources Management (ICLARM; see below), I conceived a nonparametric approach that did not require such prior knowledge or that the LF samples be weighted by abundance (or catch per effort). I also found a

Filipino programmer, Noel David, who translated the new approach, ELEFAN, into BASIC, a popular programming language at the time.[561]

ELEFAN was an instant success among colleagues working in tropical developing countries, because its various versions offered an approach for analyzing decades of accumulated LF data, using the microcomputers that were then becoming available. This led the FAO to add a comprehensive version of ELEFAN into its software series,[562,563] now simplified and rewritten in R, another programming language.[564] On the other hand, ELEFAN triggered a strong adverse reaction—centered on the non-parametric nature of the approach and the absence of formal confidence intervals. This reaction, however, was limited to colleagues in developed-country laboratories in Europe, North America, and Australia, mainly because they did not bother to inform themselves about the problems that the ELEFAN approach was supposed to solve or even about the approach itself. Similar extreme reactions were also expressed later toward Ecopath and FishBase (both of which have nevertheless become mainstream; see below). They are probably what has given me the thick skin that allows me to ignore the criticisms that my work invariably generates in some quarters, which I attribute mainly to the lack of familiarity with (or even interest in) the data-poor tropical context for which I sought solutions.

Item 2 in the above list, on understanding the basic drivers of fish growth, was also solved—at least I think so.[565,566] In fact, I think its solution is the best piece of science I have ever done, although it did not get as much attention (and hence citations) as my other work. Essentially, I found that, given the anatomical constraint that the surface area of gills represents (which determines the amount of oxygen that can be taken in by fish per unit time), the growth and final size of the fish are almost entirely determined by the ambient oxygen concentration and temperature, the latter reducing fish growth by increasing the oxygen requirements for maintenance.[567]

Gill area is a constraint, especially when the fish of a given species (or local population) become large, because in order for the fish to function, their gills have to expose a surface to the water flowing through them, and this surface area cannot grow as fast as the volume (or body mass), which must be supplied with oxygen. Many of my colleagues argue with this point, but they should not: it is a fact of geometry[568] and is also well documented empirically.[569,570,571,572,573]

The disparity between gill surface area growth (which limits the oxygen supply to the body) and body mass (which determines its oxygen requirements) elegantly explains a number of phenomena for which various *ad hoc* explanations have been proposed.[574] These phenomena[575] include:

- Why fish and aquatic invertebrates grow the way they do (asymptotically, predictably, and dependent on temperature).
- Why fish reach maturity at a size that is a predictable fraction of their maximum size, even when the latter varies because of temperature or other environmental factors.
- Why young adults sometimes "skip" spawning, while old adults have long spawning seasons.
- Why the food conversion efficiency of fish and aquatic invertebrates varies with temperature and dissolved oxygen but declines with size.
- Why aerating aquaculture ponds increases the growth and food conversion efficiency of farmed fish.

- Why fish larvae have very discernible daily rings on their otoliths (and squid larvae on their statoliths[576]) and why they become less visible in adults.
- Why the muscle tissues of young fish are full of oxidative enzymes, while those of old fish contain mainly glycolytic enzymes—that is, enzymes requiring no oxygen.
- Why visceral fat is abundant in cold temperate fish exposed to strong seasonal temperature oscillations, but not in tropical and polar fish exposed to narrow ranges of temperature.
- Why fish and aquatic invertebrates are spatially distributed the way they are and how temperature shapes their seasonal migrations.

Conventionally, in physics, chemistry, or other mature sciences, when a single hypothesis explains many phenomena, including some that had not been considered when that hypothesis was first formulated, this single hypothesis is preferred over the multiple *ad hoc* hypotheses that otherwise clutter a field.[577] Not so, apparently, in fisheries science: the hypothesis in question (which, because of its various corroborations, I now think qualifies as a theory) is hardly ever used by anyone to explain patterns and results for which it clearly does better than the "local" explanations that are commonly advanced.[578] Yet it pays to consider this theory: one of the few authors to fully embrace it, Andy Bakun, used it to explain the apparent mystery of why, for example, in typical underwater movies, larger reef fish appear to peacefully swim near smaller fish, although they could easily outswim and eat them.[579]

However, this situation might be changing now that climate change–induced ocean warming manifests globally[580] and locally.[581,582] My reason for optimism is that this theory explains, without *ad hoc* hypotheses, the observed poleward displacements of fish and invertebrates,[583] their tendency to move to deeper water,[584,585] and the tendency for their maximum size to decline,[586] all of which will have a large impact on fisheries.[587]

THE ICLARM YEARS

ICLARM, INITIALLY A Hawaii-based project of the Rockefeller Foundation, established itself in Manila, the Philippines, in 1977, under the leadership of the late Jack Marr. He had written about the demise of the California sardine fishery, coordinated the research leading to a famous book,[588] and acquired a deep knowledge of tropical fisheries while chair of the Indo-Pacific Fisheries Council. He later launched a massive FAO field project, the Indian Ocean Programme.[589]

Jack Marr had heard about my work in Indonesia and offered me a three-month consultancy in the summer of 1978 to identify researchable issues on tropical marine fisheries for ICLARM. That consultancy was a real challenge, but to my great relief, I managed to draft a coherent account. Marr sent it to a dozen leading fisheries scientists for review, including David Cushing and Brian Rothschild. To my surprise, it came back with positive comments and was published with the title *Theory and Management of Tropical Multispecies Stocks: A Review, with Emphasis on the Southeast Asian Demersal Fisheries*,[590] was reviewed positively,[591] and received hundreds of citations. It was a research program and, as it turned out, my ticket for a postdoc at ICLARM.

Jack Marr had already left ICLARM when, in the spring of 1979, I showed up again in Manila, shortly after my final doctoral examination. His successor, Ziad Shehadeh, a good-hearted Palestinian aquaculture expert, let me work there as a postdoc to implement the research program that I had defined, and after a few months he promoted me to a permanent position.

The 1980s then passed in a blur as I organized conferences, established courses and workshops in fish population dynamics run on five continents and in four languages, and published a multitude of papers, books, and reports. The many topics I covered then, which also included aquaculture,[592] were reviewed by colleagues in an edited book published in 2011.[593] It now makes me dizzy just thinking about that.

Three themes emerge from this period.

THE MULTISPECIES FISHERIES OF SAN MIGUEL BAY, THE PHILIPPINES

THE FISHERIES OF San Miguel Bay, in Bicol Province, in the northeast of the Philippines, were the first multispecies fisheries that I studied in depth. (The first fishery I studied, in Sakumo Lagoon, in Ghana, was essentially a single-species fishery for blackchin tilapia, *Sarotherodon melanotheron*.)[594] The systematic study of the San Miguel Bay fisheries, by a group of ICLARM scientists, colleagues from the University of the Philippines, and—crucially—dedicated local research assistants, who spoke Bikolano, was truly multidisciplinary from the outset. Thus, our research assistants, when interviewing fishers in the villages surrounding San Miguel Bay, could ask not only about their catch but also about their income, their expenses, and even their dreams for their children—a level of detail possible only because the research assistants grew up in these same villages.

The data recorded over a one-year cycle were analyzed by fisheries biologists,[595] fisheries economists,[596] and a rural sociologist who also lived on-site for over a year[597,598] and led to solid management recommendations for the Philippine government.[599] We also made them available to the local population via translation into two local languages.[600]

Our main result was that, while the fisheries in the bay had strongly reduced the multispecies fish assemblage, they had remained largely profitable. However, the overwhelming majority of the profits went to a few fleet owners who exploited a loophole in Philippine law that enabled them (and still does) to deploy so-called baby-trawlers in fishing grounds reserved for small-scale fishers, resulting in widespread misery around the bay.

This inequity is still not resolved, either in the Philippines or anywhere else in the world, and not only in fisheries.[601] However, I am pleased that this multidisciplinary, in-depth study of the fisheries of a remote bay in the Philippines led to the knowledge, early in my career, that stock assessments alone do not lead to an understanding of what makes fisheries tick. I also learned that when studying how natural resources are captured, one invariably hits political barriers protecting entrenched interests.

THE PERUVIAN ANCHOVETA FISHERY

FOLLOWING AN INVITATION by Wolf Arntz (also a former student of G. Hempel), who was working at the time for GTZ in Peru, I applied—with the help of Maria-Lourdes "Deng" Palomares, then my research assistant and a student, now a colleague—the ELEFAN approach to thirty years of monthly LF data on the Peruvian anchoveta (*Engraulis ringens*). Then we combined the fisheries catches and consumption by the predators of this small fish with a length-based form of Virtual Population Analysis (*en vogue* in the 1980s) to reconstruct monthly biomasses as a function of the anchoveta's rate of natural mortality (M). Given that a time-series of (fisheries-) independent acoustic estimates of anchoveta population sizes was also available, and that the anchoveta consumption by various predators (seabirds, marine mammals, and larger fish) could be computed, this indirect method allowed for monthly estimates of M, and their segmentation into predator-specific components. In a review,[602] David Cushing, then a leader in fisheries science, called the book in which these results appeared[603] "a formidable collection of papers" and a "triumph for Dr. Pauly." Who am I to disagree?

ECOSYSTEM MODELING AND THE DEVELOPMENT OF ECOPATH

AT LEAST SINCE the work of Raymond Lindeman,[604] there has been agreement that representation of aquatic ecosystems emphasizing one or the other aspects of their structure and functioning is useful for their understanding.[605] With Odum,[606] we also began to understand their patterns across wide ranges of conditions. I have long been intrigued by ecosystem models; my master's thesis, on the ecology of a Ghanaian coastal lagoon, included a simple graphical model, which inventoried the main players in that lagoon and made their ecological roles explicit.[607] Some pioneers began working on vast ecosystem models in the 1970s, but they were far too elaborate to be widely adopted, especially in the tropics. Also, at the time, the data requirements of these models could be met only in a few well-studied areas of the world, such as the North Sea[608] and the North Pacific.[609]

Thus, when I found out about the Ecopath approach and software for constructing ecosystem models that had been developed at the U.S. National Oceanic and Atmospheric Administration (NOAA)[610] and learned that its developer, Jeffrey Polovina, was not planning to work further on it, I adopted it and incorporated ideas of Robert Ulanowicz[611] and other theoreticians into it. After I presented a test case of the expanded Ecopath in Kuwait in 1987 and it received John Gulland's blessing,[612] I undertook to disseminate it widely. I was aided by Villy Christensen, who joined ICLARM in 1990 from Denmark's Fisheries and Marine Research Laboratory. Villy then led the work that established Ecopath as the most widely used approach/software for the rigorous description of aquatic ecosystem modeling.[613,614,615,616,617]

The key reason for this success, including in developing countries, was that Ecopath, like ELEFAN, could be implemented with the personal computers that had become ubiquitous. Moreover, they could easily be parameterized using widely available data—for example, the multiple diet composition studies that had so far seemed of little use—complemented by empirical relationships[618] to estimate the consumption per biomass of fish populations.

Later, Ecopath became even more popular, when a colleague at the University of British Columbia (UBC)

Fisheries Centre, Carl Walters, discovered that the easy-to-parameterize system of linear equations expressing the relationship between its state variables could be straightforwardly re-expressed as a system of differential equations specifying a time-dynamic simulation model.[619] This new routine was called Ecosim, and the integrated software was Ecopath with Ecosim or EwE. A spatial version of EwE was soon added.[620,621,622] The package, incidentally, enabled an update of my first model of Sakumo Lagoon, the Ghanaian lagoon where I did my very first research.[623]

While EwE's career continues,[624] a mid-career evaluation was provided by NOAA, which considered it one of the top ten research achievements in its 200-year history.[625]

THE 1990s: FISHBASE, INFLUENTIAL PAPERS, AND UBC

BESIDES A PUBLISHED compilation,[626] my earlier comparative work on fish growth yielded what we would now call a database. This was in the form of notecards in a wooden box, not usable by anyone else. Thus, I proposed in ICLARM's first five-year plan[627] that a widely accessible database be created:

> The information gap [presently hobbling] tropical fisheries probably cannot be bridged using [only] classical means, such as maintaining extensive libraries, encouraging interlibrary loans and electronic data exchange. Rather it can be expected that shortage of funds for such classic activities will become increasingly problematic, and hence increase the isolation of scientists working on tropical resources from the mainstream

of their science and from reference materials.... It is proposed to alleviate this problem by developing a self-sufficient database implemented on standard microcomputers... which would provide key facts and information extracted from the literature.... These facts and information will include species identification keys, morphometric data, a summary of growth and mortality information for each species, and a summary of biological data on each species. Initially, data on about 200 major species will be provided on diskettes, with the ultimate goal of 2,500 species.

I consulted with Rainer Froese, who then joined ICLARM to implement these ideas, which, the reader will realize, were a first vision of what later was to become FishBase, the online encyclopedia of fish (http://www.fishbase.org). His first suggestion was that the database should cover not only commercial fish but all fish—that is, the 20,000 species that were then thought to exist.

Then we went to work building a database, table by table, that could accommodate key information about fish, with emphasis on encoding contents, rather than trying to dazzle with a fancy interface.[628] Also, we did not implement various well-meaning suggestions that would have sunk FishBase (such as dumbing it down, requiring registration from users, and going commercial). Rather, we allowed the expansion of its contents to make it attractive and useful for an ever-widening range of user groups, with quantity slowly converting into quality.

Thanks mainly to Dr. Cornelia Nauen, also a former student of Professor Hempel, who at the time worked at the European Commission's Directorate-General for

International Cooperation and Development in Brussels, we secured a succession of large grants to implement FishBase and make it available to fisheries managers in the Commission's ACP (Africa-Caribbean-Pacific) partner countries, through training courses and annually updated CD-ROMs.[629] As a result, FishBase was well known among fisheries scientists even before it came online in 1996. Its online version also became available to a wide public interested in the over 33,000 species of fish now included, as illustrated by 50 million "hits" per month, from 0.3–0.5 million unique users, well beyond our expectations.

Because of the success of FishBase, and because many users asked us to extend it to nonfish marine organisms, we offered our database design and the special software we had developed for quicker coding and automatic verification of entries to anyone interested, but, disappointingly, there were no takers. Therefore, in 2006, with the support of the Oak Foundation, and later the Marisla Foundation, we developed SeaLifeBase (http://www.sealifebase.org), which is similar to FishBase but covers marine tetrapods (i.e., marine mammals, reptiles, and seabirds) and invertebrates.[630] SeaLifeBase now covers about 75,000 marine species and has been used, jointly with FishBase, to document, for example, the marine biodiversity of many island ecosystems later designated as marine reserves.[631,632]

In the meantime, however, ICLARM was failing as a result of a massive crisis of its governance. In the late 1980s, because we were doing well, we had been invited to join the Consultative Group on International Agricultural Research (CGIAR). This is the network of then-fourteen mostly huge R&D centers, some of which had done the research behind the Green Revolution; we knew of it because one of its key members, the International Rice Research Institute (now also the FishBase host), was based near Manila.

Joining the CGIAR, with its sprawling bureaucracy and top-down, military-style management, implemented by a succession of inept leaders, proved deadly for the creative spirits that had made ICLARM the powerhouse it was. Thus, in 1994, I accepted an offer from Tony Pitcher, then-director of the Fisheries Centre at UBC, in Vancouver, Canada, to start working there, as a professor of fisheries.

This quick ascent through the academic ranks was due, in the main, to my having previously worked at ICLARM as if I had been in academia. Thus, besides publishing, I had taught courses at the University of the Philippines (where I had about two dozen master's students), at the University of Kiel (from which, in 1984, I had received the postdoctoral degree, called a "habilitation," required in most of Europe to teach at a university,[633] and where I had my first doctoral students), and at various other universities, mainly out of a sense that this was the right thing to do. I presume the multitude of ELEFAN, FishBase, and other courses I gave in various countries also helped.

The transition from Manila to Vancouver was not easy. I had a wife with a senior position at the Manila International School, two children—a son (born in Jakarta) and a daughter (born in Manila)—whose schooling was not complete, and projects, notably FishBase, which still required nurturing.[634] Thus, from 1994 to 2000, I commuted across the Pacific, spending seven months in Canada and five months in the Philippines every year.

This period also saw the first of my contributions to *Nature* and *Science* dealing with the global impact of fisheries on marine ecosystems,[635,636,637,638,639] whose origins, main features, and impact on our discipline were later reviewed separately.[640] By covering the years from 1995 to 2003, these contributions bracketed the period during which a wide swath of the lay public realized, aided by other, seminal contributions,[641,642] that the world was experiencing not a succession of isolated, unconnected collapses but a systemic crisis of fisheries.

At UBC, I initially taught ecosystem modeling and thus introduced Ecopath to my new colleague Carl Walters, whose brilliance boosted Ecopath to a level that I could never have imagined (see above). This is, to a large extent, what turned the Fisheries Centre into a global hub for marine ecosystem modeling. We shall see if this can be maintained, given that, like ICLARM (called WorldFish since it decamped to Malaysia) and many other research powerhouses, the Fisheries Centre became a victim of its own success and was taken over by a larger body (UBC's Faculty of Science), renamed the Institute for the Oceans and Fisheries, and given a broader and blander mission.

Also, besides producing syntheses of previous work,[643,644] I expanded the range of my writing with papers on marine mammals,[645] a scholarly book on Charles Darwin and fish,[646] and whimsical essays,[647] of which one, "Anecdotes and the shifting baseline syndrome of fisheries"[648] (reprinted here), was to become quite successful. Some authors suggested that it helped launch historical ecology as a distinct discipline,[649] as also suggested in books edited by J. Jackson, K.E. Alexander, and E. Sala (*Shifting Baselines: The Past and the Future of Ocean Fisheries*)[650] and by J.N. Kittinger, L. McClenachan, K.B. Gedan, and L.K. Blight (*Marine Historical Ecology in Conservation: Applying the Past* to *Manage for the Future*),[651] while the book by D. Rost (*Misunderstanding Change: Shifting Baselines and the Perception of Environmental Change as Seen from a Sociological Perspective* [in German])[652] suggests that shifting baselines might prevent us from fully realizing the extent of, and damage done by, climate change.

In my various interactions in Canada, in the mid-1990s, I experienced a fisheries community still debating the 1992 moratorium on the fishery for northern cod (*Gadus morhua*) off Newfoundland and Labrador. Not having worked on that fishery, I never intervened in these debates or in the debates on the decline of Pacific salmon off British Columbia. To me, the issue at hand seemed not so much the specific aspects of the biology of the resources in question, or of the mathematical models used to manage them, but rather that the conversation was confined to academe and government scientists, without input from civil society.

Thus, I sought collaborations with various nongovernmental organizations (NGOs), ranging from the World Wide Fund for Nature (WWF)[653] to the nascent Marine Stewardship Council (MSC)[654]—about which, however, I later changed my opinion as it increasingly certified fisheries whose "sustainability" was widely questioned.[655,656] And so, the Pew Charitable Trusts invited me to a one-day workshop held on November 10, 1997.

THE 2000s TO THE PRESENT: THE SEA AROUND US

THE PEW CHARITABLE Trusts, then a foundation (now an NGO), and more precisely Dr. Joshua Reichert, then the director of its Environment Program, was at the time looking for a partner in assessing the "health" of the oceans to provide the scientific framework for its advocacy. Five very senior U.S. scientists and I were thus invited to a small workshop at Pew's headquarters in Philadelphia. Most of the senior colleagues quibbled about the notion of marine ecosystem "health" and proposed to set up costly data acquisition systems that would need a decade to produce useful data. In contrast, I proposed that the existing marine fisheries of the world should be seen as an ongoing "monitoring program" from which we could draw insights through judicious analyses of their catches. The paper titled "Fishing down marine food webs,"[657] which implemented this approach, was under review at the time, and I knew of what I spoke. I carried the day and was invited to submit a detailed proposal, which all of its eight reviewers found wanting, with arguments ranging from the usual ("It can't be done") to the brazen ("Give us the funding; we would do a better job of it"). Dr. Reichert then took a big risk, and the Sea Around Us, named after the second book of one of my two scientific heroes (Rachel Carson; the other is obviously Charles Darwin), began in mid-1999.

The Sea Around Us was one of the first scientific projects, and certainly the largest, meant to provide a science-based context for the advocacy work of the Pew Charitable Trusts and other environmental NGOs working on oceans and fisheries. Its mission was "to investigate the impacts of fisheries on marine ecosystems, and to propose policies to mitigate these impacts."[658] Specifically, it asked six questions about the North Atlantic (and by extension the world's oceans):

1. What are the total fisheries catches from the ecosystems, including reported and unreported landings and discards at sea?
2. What are the biological impacts of these withdrawals of biomass for the remaining life in the ecosystems?
3. What would be the likely biological and economic impacts of continuing current fishing trends?
4. What were the former states of these ecosystems before the expansion of large-scale commercial fisheries?
5. How do the present ecosystems rate on a scale from "healthy" to "unhealthy"?
6. What specific changes and management measures should be implemented to avoid continued worsening of the present situation and improve the North Atlantic ecosystem's "health"?

These questions, it will be noted, are not those that fisheries scientists usually are asked to answer. In fact, fisheries scientists often think that such strategic questions are outside their purview, if only because our discipline is very applied and tactical. Thus, most of my colleagues are directly or indirectly tasked with forecasting biomass and suggesting fishing levels and quotas, often using very sophisticated models.[659]

However, environmental NGOs are not competent, or, in fact, interested, in dealing with this tactical approach. Instead, they can pose strategic questions, on behalf of society as a whole—for example, regarding the mix of fisheries and the level of protection of Nature we should have. This is one of the functions of democracy: to allow groups of citizens to argue for environments they want for themselves and their children. This is sometimes difficult to accept for fisheries scientists, who may see themselves as working for "the fishers" or "the fisheries" and who perceive any other ethos as implying bias and advocacy. However, fisheries scientists increasingly will have to accept that fishers and fishing enterprises are not the only legitimate stakeholders; the public at large has a stake as well. Moreover, members of the public are the ones who, through their taxes, finance governments' research.

In the first five years of the Sea Around Us,[660] we documented that we had provisionally answered the above six questions for the North Atlantic and were on our way to tackling the same questions for all other oceans. Major steps in this were: (1) papers in *Nature*[661] and *Science*[662] documenting the main trends in global fisheries and in particular demonstrating that the world catch, which was at the time supposed to be increasing,[663] was actually declining, a fact masked by exaggerated catch data from China,[664] (2) a book on the state of fisheries and ecosystems in the North Atlantic Ocean,[665] and (3) the demonstration that the biomass of large fish in the North Atlantic had radically declined since the birth of industrial fisheries.[666]

In this period, our "spatialization" of the "catch" database, maintained and distributed by the FAO, based on member country contributions,[667] and available through our website (http://www.seaaroundus.org), began to be used by a large number of authors and research groups around the world, leading to numerous insights and publications, notably many articles published in *Science* and *Nature*.

As noted, our publications during this period covered the six questions mentioned above, but gradually—and this tendency became stronger in the last decade—we began to realize that Question 1 ("What are the total fisheries catches from the ecosystems, including reported and unreported landings and discards at sea?") was the most important of all, because everything else, including elaborating sound management policies, depends on accurate catch data, including that of fisheries that may be illegal.[668]

We also gradually realized that the "catch" (actually "landings") data disseminated by the FAO and used more or less uncritically by all researchers working on international fisheries throughout the world are profoundly biased. This is because FAO member countries, which contribute their data on a voluntary basis,[669] often do not cover small-scale fisheries (which are generally not small),[670] do not include discarded fish (although they have been caught),[671] and do not attempt to estimate illegal and unreported catches. Rather, problematic or difficult fisheries are ignored (treated as "no data"), and their catches (which are never zero; otherwise they would not take place) are also ignored and thus set at a figure of precisely zero. These data thus do not adequately reflect catch levels and their changes, notwithstanding a "six-decade effort to catch the trend."[672]

At the very beginning of the Sea Around Us, we had made various attempts at complementing and correcting official catch statistics.[673,674,675] However, it was only later that we fully realized the extent of this bias and the need to address it in a systematic fashion.[676]

After several years of broad-based research,[677,678] we realized that the most straightforward way to address this issue and the six questions above was through bottom-up catch "reconstructions."[679] These began formally in 2004, when the Western Pacific Regional Fishery Management Council, the management body overseeing the U.S. fisheries in the Central Pacific, contracted us to reconstruct the catch of the so-called U.S. Pacific flag territories, such as Johnston and Palmyra atolls.[680,681]

Catch reconstruction consists of a set of procedures inspired by ideas such as those presented in the essay titled "On Reconstructing Catch Time Series" and later operationalized[682] in order to derive from various sources a coherent time series of likely total catches for all fisheries of a country or area, including fisheries for which no official catch statistics exist. Catch reconstructions are also the products of these procedures. The word *reconstruction* and the concept are derived from the science of historical linguistics (one of my hobbies), which "reconstructs" extinct words or languages from daughter languages.[683]

With the crucial assistance of Dirk Zeller, the 273 catch reconstructions described in 200 distinct documents were finally completed. These reconstructions pertain either to the complete Exclusive Economic Zones (EEZs) of countries or their overseas (island) territories or to parts of their EEZs and were performed either by members of the Sea Around Us (staff, graduate students, and volunteers) or by about 300 friends and colleagues who are members of a large network created over the course of my career, working with Sea Around Us team members.[684]

This allowed us to achieve a high degree of compatibility between the reconstructions, which all covered the same period (1950–2010), used the same sectoral breakdown (industrial, artisanal, subsistence, and recreational), and accounted for both landings and discards.

The sums of all these reconstructions[685] are thus an improved estimate of the world's marine catch that can be used for various policy-relevant inferences, notably on small-scale fisheries, which have so far often been neglected by policymakers. It shows that we are catching far more fish than is being reported, implying that fisheries contribute more to our food security than we knew. These reconstructions also show that the global catch, since the mid-1990s, has been declining at a fast pace. However, this is not because prudent management has imposed low quotas in some countries (a strong decline remains when the catch of countries using quota management is subtracted). Rather, this decline of the global catch is mainly the result of a globally excessive fishing effort,[686] fueled by huge capacity-enhancing subsidies.[687] Here again, limiting one's focus to a few developed countries with functioning fisheries management systems would result in misleading conclusions about the state of global fisheries.[688]

In mid-2014, after fifteen years of fruitful collaboration, we transitioned from the Pew Charitable Trusts as our main funder to the Paul G. Allen Family Foundation, which owns Seattle-based Vulcan, Inc., and uses it to backstop the philanthropic activities it funds. In our case, the backstopping was the complete redesign and rebuilding of our website (http://www.seaaroundus.org) by an energetic group of software designers and programmers so that it draws on the contents of our newly developed database of reconstructed catches.

Thus, we now can make available to scientists, fisheries managers, NGO staffers, students, and the lay public detailed catch statistics by sector, taxon, and other criteria, spatialized by countries' EEZs, large marine ecosystems, and other geographic entities. All the data we present are based on the above-mentioned reconstruction documents (also available on our website), of which over a third have to date been published in the peer-reviewed literature. The database and website redevelopment allow a seamless transition between new data that are entered at one end (for example, correction of erroneous data as pointed out by users and frequent updates) and the output—that is, maps and time-series graphs of catches or fisheries status indicators.

Regarding status indicators, we emphasize "stock-status plots"[689,690] and the trends of declining mean trophic levels that characterize fishing down marine food webs.[691] The latter phenomenon can now be shown to be ubiquitous, thanks to a routine that overcomes the masking effects of the offshore expansion of fisheries,[692] not considered by earlier critics[693] (see also: http://www.fishingdown.org).

LOOKING FORWARD: NO MORE PYRAMIDS

THE CATCH RECONSTRUCTION work described above and the efforts to make its results globally available had their highlight in the fall of 2016 when we published the *Global Atlas of Marine Fisheries: A Critical Appraisal of Catches and Ecosystem Impacts*.[694] In addition to chapters on methods and selected themes from the work of the Sea Around Us, the atlas includes 273 one-page chapters, each covering the EEZ of a country or island territory or a part of the EEZ of larger countries.

That year I turned seventy, and thus, it is fairly certain that I won't be able to initiate or coordinate any more work of a "pyramidal" nature. Let me explain. Assuming that a goal or an insight we seek corresponds to a certain "height," and that all we have to reach that height are variously shaped stone blocks, there are two basic ways of getting there: quickly piling one block onto the other and building a tower, or cooperating with lots of people and using the blocks to build a pyramid. The latter gets to the desired height slowly (because pyramids must have an obtuse top angle), but once you get there (if colleagues and funders had the patience) what you have produced is something that won't fall down easily. In the future, however, I will build (little) towers—no more pyramids.

I am particularly happy that I succeeded in helping to empower fisheries scientists and managers in tropical developing countries by disseminating tools, concepts, and databases designed for use in those parts of the world that are data-poor and that had been neglected by fisheries science and in studies that claim to be global but are not.[695] In return, I get many citations to my work from tropical developing countries, in addition to those from developed countries, where data availability turns out not to be great either.

Overall, however, I attribute the modicum of success I have had to two factors: (1) luck in having had bosses (Gotthilf Hempel, Jack Marr, Ziad Shehadeh, and Tony Pitcher)[696] who provided me with opportunities and colleagues/friends, many of whom helped construct the pyramids mentioned above, and (2) working hard on these pyramids, logging long hours, nonstop, for decades, to compensate for my inability to build elegant towers that were tall and svelte.

If I have any advice to give, it is that one should have friends[697] and work hard.

EPILOGUE: SOME GLOOM, BUT SURELY NO DOOM

WHEN DIAGNOSING THE ills that have befallen Nature—be it our oceans or any other ecosystem—one is often labeled a propagator of "gloom and doom" and then ignored. Somehow, however dire a diagnosis is—as it is now for coral reefs under global warming—we are supposed to conclude books such as this one with an optimistic note.

This optimism-on-demand can take rather ludicrous forms, as in Jared Diamond's book *Collapse: How Societies Choose to Fail or Succeed*.[698] After outlining in agonizing detail how our civilization might fold under the weight of its excesses, the book ends with various initiatives that could help avert this collapse. For marine fisheries, the only solution discussed in any depth is to buy fish certified as sustainable by the Marine Stewardship Council (MSC).[699] However, as mentioned in several essays in this book, the MSC is not going to change the way we interact with the oceans, even if all of Jared Diamond's readers were to follow his advice. The very fact that the MSC is mentioned in this context illustrates that there is, even among authors as knowledgeable as Jared Diamond, a basic misunderstanding of the gravity of the problem we face.

I will nevertheless attempt to be positive, which I can do by noting that the great majority of the ills that out-of-control fisheries have inflicted on marine ecosystems are reversible. It would take a few decades to rebuild some of the abundance that once prevailed along the coasts, but it would be possible. Moreover, measures such as reduced fishing effort (and hence reduced fishing incomes) could be funded by diverting some of the subsidies that are now wasted on increasing fishing capacity. Such measures would not be terribly expensive and would quickly pay for themselves.[700]

Similarly, according to various studies, we could combat global warming without major impact on our economies.[701] (In fact, combating global warming would be much cheaper than the resource wars that will result from not fighting against climate change.)

But then again, there are those who say that global warming is not a problem, or not their problem, preferring instead to think that their problems are caused by the "others," that is, other people not like them.

So, gloom and doom again: we either accept that there are serious problems with the way we interact with Nature in general and with marine biodiversity in particular and step back, or…

Actually, there is no need to describe what might happen to marine biodiversity if we don't change the way our economy relates to Nature, because we will have droughts, crop failures, famines, and survival challenges that will make sustainable fisheries a frivolous issue. We will be forced to change our ways, because with no change, there might be no "we."

ABBREVIATIONS AND GLOSSARY[702]

NOTE THAT THE boldface words in the definitions are also defined in this glossary.

acoustic fish-finder: A device producing sounds that are reflected by fish schools and even by a single **fish** (especially if it has a gas bladder), whose abundance can thus be estimated.

aquaculture: According to the FAO, aquaculture is the farming of aquatic organisms, including **fish**, mollusks, crustaceans, and aquatic plants, with some sort of intervention in the rearing process to enhance production, such as regular stocking, feeding, and protection from predators, among others. See also **mariculture**.

Arctic: The region of the Earth defined as north of 66° 33′ 44″ N (i.e., north of the Arctic Circle).

artisanal fishers: Small-scale fishers who catch fish predominantly to sell them; also known as small-scale commercial **fishers**. The definitions of "small-scale" and "large-scale" (i.e., industrial) fisheries differ between countries. Thus, "artisanal" corresponds with "traditional" in Malaysia, "municipal" in the Philippines, and "*petits métiers*" in France.

atoll: A low, often roughly circular island formed when coral grows on top of the sinking cone of what was once a volcano.

bait/baitfish: Fish used to catch other fish, as in pole and line fishing or in **longlining** for tuna.

bathymetry: The measured or observed water depth.

benthos: The community of organisms that live in, near, or on the bottom of a water body.

bioaccumulation: The increase in the concentration of substances such as pesticides or other persistent organic pollutants in an organism when it absorbs the pollutants at a greater rate than the rate at which these substances are lost by the organism. Thus, the longer the biological half-life of the substance, the greater the risk of chronic poisoning, even if environmental levels of the pollutant are not very high.

biodiversity: A relatively new term used to refer to the full range of living organisms, including, *inter alia*, terrestrial, marine, and other aquatic organisms, and the ecological complexes of which they are part; this includes the diversity within species, between species, and in **ecosystems**.

biomass: Weight of a "stock" or population of fish, or of one of its components; thus, for example, "spawning biomass" is the combined weight of all sexually mature animals in a population. Used as a measure of population abundance.

bycatch: That part of a fish **catch** that is caught in addition to the "target species" because the fishing gear (e.g., a **trawl**) is not selective. Bycatch may be retained, landed, and sold or used, or it may be dumped at sea (see **discard**).

capacity (carrying): The average size of the population of a given species (not exploited by humans) that can live (i.e., feed itself and survive) in a given **ecosystem**.

capacity (fleet): The minimum fleet size and **effort** required to generate a given **catch**, and as well, the maximum catch that a **fisher** or a fleet can produce with given levels of inputs, such as fuel, amount of fishing gear, ice, **bait**, engine horsepower, and vessel size.

cascade (trophic): A trophic cascade occurs when predators in a **food web** suppress the abundance or alter the behavior of their prey so that the next lower **trophic level** (i.e., the prey of the prey) is released from predation (or **herbivory** if the intermediate trophic level is a herbivore). For example, if the abundance of large piscivorous fish is increased, the abundance of their prey, zooplanktivorous fish, should decrease, large **zooplankton** abundance should increase, and **phytoplankton** biomass should decrease. This concept has stimulated research in many areas of marine ecology and fisheries biology.

catch: The number or weight of **fish** or other animals caught or killed by a fishery, including fishes that are landed (**landings**, whether reported in statistics or not), discarded at sea (**discards**), or killed by lost gear ("**ghost fishing**").

catch composition: The different taxa (species, genera, family; see **taxon**) making up the **catch** of a fishery. The more detailed a catch composition is,

catch/effort (or catch per unit of effort): A measure of relative abundance, obtained by dividing the **catch** by a measure of the fishing **effort** required to realize this catch. Catch/effort is generally proportional to **biomass**.

climate change: A lasting change in the statistical distribution of weather patterns over periods ranging from decades to millions of years. This may be a change in average weather conditions or in the distribution of weather around the average conditions (i.e., more or fewer extreme weather events such as storms or heat waves), or both. Climate change is caused by factors such as variations in solar radiation received by Earth, plate tectonics, and volcanic eruptions. The release of greenhouse gases (notably carbon dioxide and methane) by human activities has also been identified as the cause of recent climate changes, often referred to as "global warming."

cod end: The terminal part of a **trawl** net, where the **fish** and invertebrates caught end up.

collapse(d): Here: the rapid decline in the abundance (**biomass**) of a **fish** population, generally reflected in a rapid decline of the **catches** taken from that population, either because there are fewer fish to be caught than previously or, less often, because the **fishery** that exploited it is closed (as when a population is protected by a **marine reserve**) or strongly reduced (to enable **rebuilding**). Many researchers believe that unless special circumstances apply, a fish population has collapsed when the catch taken from it is below 10 percent of its historical maximum.

commercial fishery: A **fishery** whose **catch** is sold. This means that both **large-scale** (or **industrial**) and **small-scale fisheries** (or **artisanal**, or *petit métiers*) are commercial fisheries, and that the term "commercial fisheries" should not be considered synonymous with industrial or large-scale fisheries.

Common Fisheries Policy (CFP): The policy governing the marine **fisheries** in the 9.7-million-square-mile (25 million km^2) **EEZs** jointly held by the member states of the **European Union (EU)**. The CFP covers a range of activities, of which the main one is probably setting annual **quotas** (suggested by ICES) and allocating them to the fleets of its maritime member states. The CFP, whose policy outcome had been widely criticized, was reformed in 2013 through an act of the EU Parliament, with the result that more emphasis will be given to stock **rebuilding** and the gradual banning of **discards**.

continental shelf: See shelf.

demersal: Organisms swimming just above or lying on the seafloor and usually feeding on benthic organisms (the **benthos**).

discard: The portion of a **catch** that is thrown overboard but that may be of important ecological or commercial value. Discards typically consist of "nontarget" species or undersized specimens of the **target species**. High-grading is a special form of (mostly illegal) discarding where a catch of target species is thrown overboard to make space in the hull (or to accommodate under a **quota**) fresher, larger, or otherwise more valuable catch of the same species.

distant-water fleet/fishery: The fleet of a country that is fishing in the EEZ of another country (or the EEZs of other countries), or in **high sea** regions not adjacent to its own EEZ. Under the UNCLOS, a distant-water fishery can be conducted in the EEZ of a coastal state only with an explicit access agreement, generally in exchange for compensation.

domestic waters: A country's or territory's own EEZ, comprising territorial waters, extending to twelve miles off the coast, and a zone of extended jurisdiction reaching at most 200 miles off the coast.

Ecopath: An approach and software package allowing for the straightforward construction of mass–balance models (quantified representations) of the trophic linkages in aquatic **ecosystems** at a given time or during a given period.

Ecosim: An add-on to **Ecopath** that uses its parameterization to define a system of differential equations allowing changes in, for example, fishing tactics or environmental forcing to be evaluated according to their effects on the time-series dynamics of an **ecosystem** as a whole.

Ecospace: An add-on to **Ecosim** that allows the processes it simulates to be represented spatially.

ecosystem: A community of plants, animals, and other living organisms, together with the non-living components of their environment, found in a particular habitat and interacting with each other.

EEZ: See Exclusive Economic Zone.

effort (fishing): Any activity or device that is deployed to catch **fish** and that can be quantified. Thus, the number of days fished in a given time period (typically a year) is a measure of effort, as are the number of nets of a certain type deployed in a set period and the amount of fuel used by a fishing fleet in a certain period of time.

endemic species: Species native in, and restricted to, a particular area, (e.g., an island, a country, a continent, an ocean).

estuary: The brackish mouth of a river, resulting in transitional environments between freshwater and salt water.

European Union (EU): A unique political and economic union and partnership of twenty-eight states in Europe that functions through intergovernmental negotiated decisions and supranational institutions (e.g., the EU Parliament, the EU Commission). The EU encompasses over 1.5 million square miles (4 million km^2) and has over 500 million inhabitants, and as a single market it is a major world trading power. In 2016, a majority of UK voters opted for the United Kingdom to leave the twenty-seven other countries of the EU, which is a pity.

Exclusive Economic Zone (EEZ): Generally, all waters within 200 nautical miles (370 km) of a country and its outlying islands, unless such areas would overlap with neighboring countries that are less than 400 nautical miles (740 km) apart. If an overlap exists, it is up to the countries to negotiate a delineation of the actual maritime boundary. Under the UNCLOS, a country has special rights regarding the exploration and use of marine resources inside its EEZ, such as the power to control and manage all fisheries resources in this zone. Not until 1982, with the adoption of the UNCLOS, did 200-nautical-mile EEZs become formally adopted, and a country needs to formally declare its EEZ. Note that EEZs are not "territorial waters," which are the areas beyond the tidal baseline of the open coasts of a country over which that country exercises full control and sovereignty except for innocent passage of foreign vessels and which are set at a maximum of twelve nautical miles by the 1982 UNCLOS.

ex-vessel value: The price that **fishers** get for a unit weight of **fish** landed at a dock or beach multiplied by the weight of the **landings** at this first point of sale; corresponds to "farm gate" prices and value.

FADs: See **fish aggregating devices**.

FAO: See **Food and Agriculture Organization of the United Nations**.

fish: The term "fish" *sensu stricto* refers to the taxonomic (**taxon**) class Pisces in the Subphylum Vertebrata, Phylum Chordata. In the wider sense, used throughout this book, "fish" also includes the aquatic invertebrates sought by **fisheries**.

fish aggregating devices (FADs): Floating objects made of vegetable matter (e.g., palm fronds) or artificial materials (e.g., plastic, steel, and even concrete), some anchored on the seafloor, where they attract **pelagic fishes**, mainly tuna, because of their propensity to congregate under floating debris. Some FADs are now equipped with sensors (linked to satellites) to assess when they can be fished. FADs attract juvenile fish and thus can contribute to **overfishing**.

fisher: A person who fishes in any kind of **fishery**, whether small- or large-scale.

fisheries-independent data: Information about a **fishery** resource *not* based on catches and statistics derived from **catches**, such as the **catch/effort**, of fishing vessels. Typically, fisheries-independent data are obtained from dedicated research vessels performing **trawling** or acoustic surveys.

fishery: A set of persons and gear interacting with an aquatic resource (one or several species of **fish**) for the purpose of generating a **catch**.

fishing down (marine food webs): The process whereby **fisheries**, in a given **ecosystem**, having depleted the large predatory **fish** at the top of the food web, turn to increasingly smaller species, ending up with previously spurned small fish and invertebrates. See also the Wikipedia entry on this and http://www.fishingdown.org.

fishing effort: See **effort**.

fishmeal: Protein-rich animal feed product based on ground-up **fish**, usually small **pelagic fishes** such as anchovies and sardines, which are also directly consumed by people.

F$_{MSY}$: The value of fishing mortality F, which produces the maximum yield in the long term; see also **MSY**.

Food and Agriculture Organization of the United Nations (FAO): A UN technical agency based in Rome that is responsible for managing information and programs related to human food security (see http://www.fao.org). It is also the only agency in the world tasked with annually assembling global **fisheries** statistics and assisting member countries in managing their fisheries.

food chain: A hierarchical arrangement of organisms according to **trophic level**.

food security: According to the **FAO**, this occurs "when all people, at all times, have physical and economic access to sufficient, safe and nutritious food to meet their dietary needs and food preferences for an active and healthy life." Seafood contributes crucially to food security in numerous countries where alternative sources of animal protein and micronutrients are lacking.

food web: All the interacting **food chains** within a particular community.

forage fish: Small **pelagic fish**, often also called **bait fish**, which are a major food item for larger predators, such as larger **fish**, **seabirds**, and **marine mammals**. Forage fish usually feed near the base of the **food chain**, on **plankton**.

ghost fishing: The killing of **fish** by lost gears (traps, gill nets, etc.)

herbivory: The feeding adaptation of herbivores, or animals consuming foliage and grass on land. In the sea, herbivory is practiced by **zooplankton** feeding on **phytoplankton** (i.e., microscopic algae), and by **fishes**, such as sardines, also capable of filtering phytoplankton. Herbivorous fishes such as parrotfish and invertebrates such as sea urchins also consume fleshy algae or kelp.

high-grading: See **discards**.

high sea(s): The areas of the world ocean that are outside the 200-nautical-mile (370-km) **Exclusive Economic Zones** of coastal states; the high seas cover about 60 percent of the world's oceans.

ICES: See **International Council for the Exploration of the Sea**.

illegal fishing: Fishing in violation of the laws of a fishery, either under the jurisdiction of a coastal state (i.e., within an EEZ) or in high seas fisheries regulated by **Regional Fisheries Management Organizations (RFMO)**. Here, "illegal" is defined as fishing in the EEZ waters of another country without explicit or implicit permission or an "access agreement" and thus does not include noncompliance with domestic fishing laws by domestic fleets in their own EEZ (i.e., poaching).

individual transferable quota (ITQ): An ITQ is a type of catch share used by many governments to regulate commercial fishing. The regulatory agency sets a species-specific **total allowable catch (TAC)**, typically by weight and for a certain time period (e.g., annually). A percentage of the TAC, called quota shares, is then allocated to individual **fishers** or fishing entities (e.g., companies, communities). An ITQ is transferable, meaning that it can typically be sold and bought. Thus ITQs often end up being concentrated in the hands of financial operators (e.g., banks, hedge funds), and the fishers become hired (and fired) hands.

industrial fisheries: Fisheries that catch fish for commercial marketing or global export (i.e., large-scale commercial fisheries). The distinction between large-scale and **small-scale** commercial (also known as **artisanal**) fisheries often differs between countries and is usually related to vessel size and gear type used.

International Council for the Exploration of the Sea (ICES): Founded in 1902, ICES is the oldest intergovernmental organization in the world concerned with marine and fisheries science. ICES conducts research and provides advice to manage fisheries in the waters of the countries of the European Union and associated partner countries. ICES is headquartered in Copenhagen and consists of a network of over 350 marine institutes in twenty member countries and beyond. Its work also extends into the Arctic, the Mediterranean Sea, the Black Sea, and the North Pacific Ocean.

ITQ: See **individual transferable quota**.

IUU: Illegal, unreported, and unregulated; an abbreviation proposed by the **FAO** to describe problematic **fisheries** and their **catches**. This abbreviation is not used much in this book, because it has become synonymous with "illegal" in practice and thus confuses people.

lagoon: A small water body associated with either coastlines or coral reefs. Coastal lagoons are formed behind permanent or occasionally (or seasonally) occurring sand bars, and their salinity (and hence their flora and fauna) depends on their water exchange with rivers and the sea (through breaks in the sand bars). Coastal lagoons can be very productive, as, for example, along the coast of the Gulf of Guinea. The shallow water body whose periphery is defined by **atolls** is also called a lagoon, and its productivity, if it is open to the outside, can also be high. Both lagoon types are vulnerable to sudden drops in salinity, which can kill the resident organisms.

landing: The weight of the **catch** landed at a wharf or beach. Also: the number or weight of **fish** unloaded at a dock by **commercial fishers** or brought to shore by **subsistence** and **recreational fishers** for personal use. Landings are reported at the points at which fish are brought to shore. Note that the catch equals the fish landed plus **discards**.

large marine ecosystem (LME): A large area of ocean space (often about 77,000 square miles or 200,000 km^2, or greater) adjacent to the continents in coastal waters, where primary production (see **production**) is generally higher than in open ocean areas. The delineation and definition of individual LMEs and their boundaries are based on four ecological, rather than political (see **EEZ**) or economic, criteria. These are (1) bathymetry, (2) hydrography, (3) productivity, and (4) trophic relationships. Based on these ecological criteria, sixty-six distinct LMEs have been delineated to date around the coastal margins of the Atlantic, Pacific, and Indian Oceans.

large-scale fishery: See **industrial fisheries**.

LME: See **large marine ecosystem**.

longlining: Fishing by means of a line that can be a mile or more (several kilometers) long and that bears numerous baited hooks, usually set horizontally in the water column and used, for example, in snapper or grouper and especially tuna **fisheries**. The line is set for varying periods up to several hours on the seafloor, or in the case of tuna, in mid-water at various depths. Also refers to a fishing line that has baited hooks set at intervals on branch lines and that may be 90 miles (150 km) long and have hundreds of thousands of hooks; it can be on the sea bed or above it, supported by floats. Such longlines may be anchored or drift free and are marked by floats.

mangrove: Trees and shrubs that grow in saline coastal sediment habitats in the tropics and subtropics, mainly between latitudes 25° N and 25° S. Mangroves are saline wood- or shrub-land habitats that are characterized by depositional coastal environments, where fine sediments (often with high organic content) collect in areas protected from high-energy wave action. The saline conditions tolerated by various mangrove species range from brackish water to pure seawater to water concentrated by evaporation to over twice the salinity of ocean seawater. Mangroves are crucial juvenile nursery habitats for many **fisheries** species and also, like healthy coral reefs, shield coastlines from the impacts of high-energy ocean surges and damage from tropical storms.

mariculture: The farming of aquatic organisms in seawater, such as fjords and inshore and open waters, in which the salinity generally exceeds 20 percent. These organisms may spend earlier stages of their life cycle in brackish water or freshwater.

marine mammals: A diverse group of well over 100 species that rely on the ocean for their existence, including seals, whales, dolphins, porpoises, manatees, dugongs, otters, walruses, and polar bears. They do not represent a distinct biological grouping but rather are unified by their reliance on the aquatic environment for feeding, although their dependence on aquatic **ecosystems** varies considerably between species. Marine mammals can be subdivided into four major groups: cetaceans (whales, dolphins, and porpoises), pinnipeds (seals, sea lions, and walruses), sirenians (manatees and dugongs), and fissipeds, which are the group of carnivores with separate digits (polar bears and two species of otters). Both cetaceans and sirenians are fully and obligate ocean dwellers. Pinnipeds are semiaquatic, meaning that they spend most of their time in the water but need to return to land for mating, breeding, and molting. Otters and polar bears are much less modified for ocean living. While the number of marine mammals is small compared with that of land mammals, their total **biomass** is large. The hunting of marine mammals has both **artisanal/subsistence** and (highly destructive) industrial components.

marine protected area (MPA): An area of the ocean within which fishing and other extractive activities are limited. Often used to mean "**no-take area**," where no fishing is allowed, but the term "**marine reserve**" is more appropriate for such an area.

marine reserve: A form of **marine protected area** that includes legal protection against fishing (i.e., is designated a "no-take area"), meaning that it is an area of ocean space where all fishing is prohibited. Benefits include increases in the biodiversity, abundance, **biomass**, body size, and reproductive output of **fisheries** populations.

marine trophic index (MTI): The mean **trophic level** of fisheries **landings**.

maximum sustainable yield (MSY): The maximum amount that can be taken (caught) over the long term from a **fishery**. MSY is now considered an upper limit for fishery management as opposed to a **target level**.

MPA: See **marine protected area**.

MSY: See **maximum sustainable yield**.

natural mortality: A mathematical expression, usually represented by the letter M, of the part of the total rate of deaths of **fish** from all causes except fishing (e.g., predation, cannibalism, disease, and senescence).

neritic zone: The shallow part of the **pelagic zone** that reaches over the continental **shelf**, down to a depth of 650 feet (200 meters).

no-take area: A **marine reserve** in which no fishing is allowed.

nutrient: A substance required by organisms for body health and growth; for marine **primary production**, the key nutrients are nitrates, silicates, and phosphates, generally supplied by tidal and wind mixing and **upwellings**.

overfishing: Applying a level of fishing **effort** beyond that which will generate a desirable, **sustainable**, or "safe" population level. The level of effort can be in excess of that required to generate **maximum sustainable yield** (biological overfishing), maximum economic yield (economic overfishing), maximum yield per recruit (growth overfishing), or maximum **recruitment** (recruitment overfishing). See also the discussion of Malthusian overfishing, in the essay titled "Major Trends in Small-Scale Fisheries."

pelagic fish: Fish that live and feed in the open sea; associated with the surface or middle depths of a body of water; free-swimming fish in the seas, oceans, or open waters, but not in association with the bottom (**benthos**). Most pelagic fish feed on **plankton**.

pelagic zone: The zone of open waters in lakes, seas, and oceans. The pelagic zone is often considered to reach from the water surface to depths of 1,600–3,300 feet (500–1,000 meters), at which point the mesopelagic zone begins.

phytoplankton: The smallest components of the **plankton** community, consisting of microscopic plants, which are a key component of marine and freshwater ecosystems. They are the major transformer of solar energy into **biomass** that "feeds" aquatic **food webs**.

plankton: The community of living plants (microscopic **phytoplankton**) and animals (**zooplankton**) whose lack of powerful propulsive organs leaves them to drift with the water body in which the vagaries of turbulence, currents, or **upwellings** have placed them.

population: A set of interacting organisms of the same species that live in the same geographical area and can interbreed. Roughly corresponds to the concept of **stock** used by fisheries scientists.

primary production: The process by which plants (including **phytoplankton** and other algae) use light energy to transform carbon dioxide and water into vegetable biomass that can then be grazed by **herbivores**.

producers: See **production, primary production, phytoplankton,** and **zooplankton**.

production: In ecology and **fisheries** biology, the sum of all growth increments of the animals or plants of a **population** over a given time period, including the growth of individuals that may not have survived to the end of that period. Most of the **primary production** of the ocean is due to **phytoplankton**, while secondary production is due to herbivorous **zooplankton**. The term "production," which is appropriate in agriculture, may be applied to **aquaculture** (including **mariculture**) but should never be applied to fisheries **catches** (or even **landings**), which are not "produced" by fishing.

purse seine: A fishing net used to encircle and catch **pelagic fish**. The net may be up to two-thirds of a mile (1 km) long and 1,000 feet (300 m) deep and is used to encircle surface schooling fish such as anchovies, mackerel, or tuna. One end of the net is usually set at speed from a larger vessel, while the other end is anchored by a small boat. During retrieval, the bottom of the net is closed, or pursed, by drawing a purse line through a series of rings to prevent the fish from escaping.

quota: The amount of fish that a country, enterprise, or **fisher** is allowed to take in a given year. Also refers to a constant fraction of a variable **total allowable catch** (TAC).

rebuilding: Reducing fishing (e.g., via low to zero **quotas**) until the natural processes of **recruitment** and individual growth cause the **biomass** of a **stock** (or **population**) to increase to some pre-set level (e.g., that generating **maximum sustainable yield** [MSY])

reconstruction (catch): The set of procedures used to derive a coherent series of likely annual total **catches** for all **fisheries** of a country or area from various sources, not necessarily including official catch statistics; also, the product of these procedures. The word and concept are derived from the science of linguistics, which "reconstructs" extinct words (and languages) from daughter languages.

recreational fishing: Fishing for pleasure (also called sport fishing), in contrast to **commercial fishing** (both **artisanal** and **industrial**), where the main motivation is to catch **fish** for eventual sale, and from **subsistence**

fishing, where fish is mainly caught for personal consumption or consumption by family and friends.

recruitment: The process by which young **fish** enter a fish **population**. Recruitment is distinguished from the term "reproduction" because the high mortality experienced by fish eggs and larvae usually precludes the prediction of population sizes from the abundance of these early stages (while the number of recruits can be used to predict the number of adults).

reduction fisheries: Fisheries whose **catch** is used for making **fishmeal** (often with fish oil as a by-product), which is then used to feed animals (e.g., pigs, chickens, or farm-raised salmon). An example of a **fish** that is mostly "reduced" to fishmeal is the Peruvian anchoveta (*Engraulis ringens*). Most fish used for fishmeal are perfectly edible, and thus reduction fisheries are usually competing with fisheries whose catch is used for direct human consumption.

Regional Fisheries Management Organizations: International governmental organizations tasked with managing the **fisheries** of a region of the ocean (including the **high sea**) for the benefit of member states.

Regional Fishery Management Councils: The eight bodies in the United States that are responsible for the management of **fisheries** in the U.S. **Exclusive Economic Zone**. The councils, established by the Magnuson–Stevens Fishery Conservation and Management Act in 1976, include federal and state officials and at-large and obligatory members selected by state governors to represent non-government stakeholders and special interest groups such as commercial **fishers**.

resilience: The capacity of a system to tolerate negative impacts without irreversible change in its outputs or structure. In a species or **population**, resilience is usually understood as the capacity to withstand exploitation.

RFMC: See **Regional Fishery Management Councils**.

RFMO: See **Regional Fisheries Management Organizations**.

Sea Around Us, the: The name of the research project led by the author at the Fisheries Centre (now Institute for the Oceans and Fisheries) of the University of British Columbia since July 1999, many of whose results are presented in this book (see also http://www.seaaroundus.org). The Sea Around Us was named after the 1951 bestselling book by Rachel Carson.

seabirds: Birds that have adapted to life within the marine environment, although they vary greatly in behavior and physiology. Seabirds usually have a longer lifespan, breed later, and have fewer young than other types of birds, but they invest a great deal of time in their young. Most seabirds nest in colonies and may make long annual migrations.

shelf: The seafloor between the coast and the 650-foot (200 m) isobath around the continents (the continental shelf) and, less commonly, around islands. Shelves are the most productive parts of the oceans and support their most important **fisheries**.

small-scale fishery: Artisanal, subsistence, and recreational fisheries.

sport fishing: See recreational fishing.

stock: A term quasi-synonymous with "**population**," commonly used by **fisheries** scientists; in the narrow sense, "stock" refers to the exploited part of a **fish** population.

stock assessment: A set of mathematical procedures using data from life-history studies, **fisheries**-independent surveys, and **catch** statistics to estimate the current and probable future abundance (or **biomass**) of commercial fish **stocks**. Generally forms the basis for setting **total allowable catches**.

stock-status plot: A manner of summarizing time series of fisheries catches so that trends in the status of the **fisheries** that generated these **catches** are highlighted (see http://www.seaaroundus.org).

subsidies: Government funds made available to a segment of the population of a country or a sector of its economy. When given to a well-developed **fishery**, subsidies tend to encourage **overfishing**.

subsistence fishing: A form of small-scale fishing often practiced by women and children (e.g., as "gleaning" on reef flats), where the **catch** (often small **fish** and invertebrates, particularly bivalves) is mainly caught for personal consumption or consumption by family members or is bartered against other commodities.

sustainability: The ability of an activity or process to be maintained indefinitely (or at least into the long-term future). "Sustainable growth," by this definition, is an oxymoron.

sustainable: See **sustainability**.

TAC: See **total allowable catch**.

target: This term has two meanings in fisheries. One refers to the **fish** (species or group) that are meant to be caught, as opposed to **bycatch**, which is caught because the gear targeting a certain type of fish is not selective or is insufficiently selective. The other meaning refers to **fisheries** management, which often defines 90 percent of MSY as the targeted catch, while MSY itself is a limit never to be exceeded.

taxon (plural: taxa): According to the International Code of Zoological Nomenclature, any formal unit or category of related organisms (species, genus, family, order, class, etc.). Derived terms are *taxonomist, taxonomic, taxonomically*.

threatened species: Species of animals, plants, fungi, etc., that are vulnerable to endangerment. The International Union for Conservation of Nature (IUCN) is the foremost authority on threatened species and identifies three categories of threatened species (in their "Red List"), depending on the degree to which they are threatened: Vulnerable species, Endangered species, and Critically Endangered species. Less-than-threatened categories are Near Threatened, and Least Concern. Species that have not been evaluated (NE) or for which there is not sufficient data (Data Deficient) are not considered threatened by the IUCN.

ton (metric): A weight unit, corresponding to a "tonne," or 1,000 kilograms (2,200 pounds).

total allowable catch (TAC): The amount of **fish** (or **quota**) that can be taken legally by a given **fishery** in a given period (usually a year or a fishing season), as determined by a fishery management body, such as the Department of Fisheries and Oceans in Canada.

traditional fisher or fishery: Terms misleadingly used in some countries to describe **artisanal fishers** and **fisheries**.

trash fish: The earlier and badly misleading name for the fraction of the **bycatch** for which no market had been identified and which was therefore discarded.

trawl: A fishing method in which a vessel—a trawler—tows a large bag-shaped trawl net. A wide range of bottom (also called **demersal** or benthic) or **pelagic** (open water) species of **fish** are taken by this fishing method. The trawl net usually features a buoyed head (top) rope, a weighted foot (bottom) rope, and two "otter" doors to keep the mouth of the net open. Variations include beam trawling, which uses a horizontal beam instead of otter doors and foot rope to keep the net open, and pair-trawling, in which two vessels are used to tow a single, often huge net. Bottom trawling is unselective and destructive of habitats and is gradually being banned from areas that people care about. All trawls are here considered industrial gear, whatever the size of the vessel pulling them.

trophic level: Numbers expressing the relative "height" of an organism within the food web, with plants having a trophic level (TL) of 1, herbivores 2, their predators 3, and so on. Because **fishes** have mixed diets, they tend to have intermediate TL values (e.g., 2.5 for an omnivore feeding half on plants and half on herbivores).

tropics: A climate zone ranging north and south from the equator to the limits of the subtropical zone and generally limited to sea surface temperatures above 20° C.

UNCLOS: See United Nations Convention on the Law of the Sea.

United Nations Convention on the Law of the Sea (UNCLOS): Also called the Law of the Sea Convention or the Law of the Sea Treaty, this is the international agreement that defines the rights and responsibilities of nations with respect to their use of the world's oceans, establishing guidelines for businesses, the environment, and the management of marine natural resources. Among other things, the UNCLOS has enabled countries to declare an EEZ to a maximum of 200 nautical miles (370 km).

upwelling: An oceanographic phenomenon involving the wind-driven rise of dense, cooler, and usually nutrient-rich water toward the ocean surface, where it replaces (and pushes offshore) warmer, usually nutrient-depleted surface water. The cold but nutrient-rich upwelled water stimulates the growth of primary **producers** (mainly **phytoplankton**) and secondary producers (mainly **zooplankton**), upon which the other inhabitants of the **ecosystem** (**forage fish**, other **fishes**, and **marine mammals**) depend.

yield: Catch in weight during a conventional period (e.g., a year); see also **maximum sustainable yield (MSY)**.

zooplankton: Small components of the **plankton** community, consisting of small to microscopic animals, which feed on either **phytoplankton** or other zooplankton. They form part of the bottom of the **food chain** and are often a major food source for small- to medium-sized **fishes** and invertebrates.

ENDNOTES

THE FOLLOWING NOTES consist of the original footnotes, endnotes, and references for the contributions reprinted in this book or, when preceded by N.N. (for New Note), of the notes that were added to provide context or updates for this edition.

[1] Jackson, J.B.C., M.X. Kirby, W.H. Berger, K.A. Bjorndal, L.W. Botsford, B.J. Bourque, R. Cooke, J.A. Estes, T.P. Hughes, S. Kidwell, C.B. Lange, H.S. Lenihan, J.M. Pandolfi, C.H. Peterson, R.S. Steneck, M.J. Tegner, and R.R. Warner. 2001. "Historical overfishing and the recent collapse of coastal ecosystems." *Science* 293: 629–638.

[2] N.N. This contribution, reprinted here with permission, was originally published as Pauly, D. 2009. "Beyond duplicity and ignorance in global fisheries." *Scientia Marina* 73(2): 215–223. It acknowledged the Honorable Mr. José Montilla, President of the Generalitat of Catalonia (Spain), and the Management and Jury of the Ramon Margalef Prize for awarding me this prize. I thus had the opportunity to prepare the initial version of this essay, which, although concerned mainly with fisheries management, also deals with aquatic ecosystems. Ramon Margalef studied mainly the lower trophic levels of aquatic ecosystems, so with my account here, covering mainly their upper trophic level, we have them covered, as the phrase goes, "from end to end." My gratitude also goes to Drs. Marta Coll, Isabel Palomera, and John Celecia and to Ms. Teresa Sala Rovira, who all contributed to my sojourn in Catalonia, and to Dr. M.P. Olivar for explicitly inviting the original version of this essay, and suggesting its title.

³ N.N. Obviously, *national* catch statistics existed long before 1930, while statistics for specific fisheries can go back centuries; see, e.g.: Ravier, C. and J.M. Fromentin. 2001. "Long-term fluctuations in the eastern Atlantic and Mediterranean bluefin tuna population." ICES Journal of Marine Science 58: 1299–1317.

⁴ Ward, M. 2004. *Quantifying the World: UN Ideas and Statistics.* Bloomington: Indiana University Press.

⁵ N.N. Every two years, the FAO also publishes extremely valuable analyses of trends in fisheries data, in the form of a report called the *State of Fisheries and Aquaculture*, or SOFIA, available from their website. For comments on SOFIA 2016, see: Pauly, D. and D. Zeller. 2017. "Comments on FAOs State of World Fisheries and Aquaculture (SOFIA 2016)." Marine Policy 77: 176–181.

⁶ This observation is based on experience teaching fisheries science in four languages on five continents and interacting with hundreds of colleagues, but with a bias toward developing countries.

⁷ N.N. In spite of the following paper's title, the issue of missing catches and their distorting effect on long-term trends was not addressed in Garibaldi, L. 2012. "The FAO global capture production database: A six-decade effort to catch the trend." Marine Policy 36: 760–768.

⁸ N.N. For trends in cumulative engine power, see: Anticamara, J.A., R. Watson, A. Gelchu, and D. Pauly. 2011. "Global fishing effort (1950–2010): Trends, gaps, and implications." Fisheries Research 107: 131–136, and Watson, R., W.W.L. Cheung, J. Anticamara, U.R. Sumaila, D. Zeller, and D. Pauly. 2013. "Global marine yield halved as fishing intensity redoubles." Fish and Fisheries 14: 493–503.

⁹ N.N. See: http://www.seaaroundus.org for such data, covering the years 1950 to 2004 at various scales. E.g., Countries' EEZs, Large Marine Ecosystems, FAO Major Fishing Areas.

¹⁰ Radovich, J. 1981. "The collapse of the California sardine industry: What have we learned?" In: *Resource Management and Environmental Uncertainty*, edited by M.H. Glantz and D. Thomson, 107–136. New York: Wiley.

¹¹ Beverton, R.J.H. 1990. "Small pelagic fish and the threat of fishing: Are they threatened?" Journal of Fish Biology (Suppl. A): 5–16.

¹² Muck, P. 1989. "Major trends in the pelagic ecosystem off Peru and their implications for management." In: *The Peruvian Upwelling Ecosystem: Dynamics and Interactions*, edited by D. Pauly, P. Muck, J. Mendo, and I. Tsukayama, 386–403. ICLARM Conference Proceedings 18. Manila: International Center for Living Aquatic Resources Management.

[13] Castillo, S. and J. Mendo. 1987. "Estimation of unregistered Peruvian anchoveta (*Engraulis ringens*) in official catch statistics, 1951 to 1982." In: *The Peruvian Anchoveta and its Upwelling Ecosystem: Three Decades of Changes*, edited by D. Pauly and I. Tsukayama, 109–116. ICLARM Studies and Reviews 15. Manila: International Center for Living Aquatic Resources Management.

[14] N.N. The Castillo/Mendo estimate was confirmed by a subsequent study available on http://www.seaaroundus.org (see: Peru), and summarized in Mendo, J. and C. Wosnitza-Mendo. 2016. "Peru." In: *Global Atlas of Marine Fisheries: A Critical Appraisal of Catches and Ecosystem Impacts*, edited by D. Pauly and D. Zeller, 366. Washington, DC: Island Press.

[15] Hardin, G. 1968. "The tragedy of the commons." *Science* 162: 1243–1248.

[16] Pauly, D. 2007. "On bycatch, or how W.H.L. Allsopp coined a new word and created new insights." *Sea Around Us Project Newsletter* 44: 1–4.

[17] Pauly, D. and J. Maclean. 2003. *In a Perfect Ocean: The State of Fisheries and Ecosystems in the North Atlantic Ocean*. Washington, DC: Island Press.

[18] Rose, A. 2008. *Who Killed the Grand Banks: The Untold Story Behind the Decimation of One of the World's Greatest Natural Resources*. Mississauga, ON: John Wiley and Sons.

[19] Jackson, J.B.C., M.X. Kirby, W.H. Berger, K.A. Bjorndal, L.W. Botsford, B.J. Bourque, R. Cooke, J.A. Estes, T.P. Hughes, S. Kidwell, C.B. Lange, H.S. Lenihan, J.M. Pandolfi, C.H. Peterson, R.S. Steneck, M.J. Tegner, and R.R. Warner. 2001. "Historical overfishing and the recent collapse of coastal ecosystems." *Science* 293: 629-638.

[20] Roberts, C. 2007. *The Unnatural History of the Sea*. Washington, DC: Island Press.

[21] Bonfil, R., G. Munro, U.R. Sumaila, H. Valtysson, M. Wright, T. Pitcher, D. Preikshot, N. Haggan, and D. Pauly. 1998. "Impacts of distant water fleets: An ecological, economic and social assessment." In: *The Footprint of Distant Water Fleets on World Fisheries*, Endangered Seas Campaign, WWF International, 11–111. Godalming, Surrey: WWF International. (Also issued separately as Bonfil et al., eds. 1998. "The footprint of distant water fleets on world fisheries." *Fisheries Centre Research Reports* 6(6). Vancouver, BC: University of British Columbia.)

[22] Pauly, D. 1986. "Problems of tropical inshore fisheries: Fishery research on tropical soft-bottom communities and the evolution of its conceptual base." In: *Ocean Yearbook 1986*, edited by E.M. Borgese and N. Ginsburg, 29–37. Chicago: University of Chicago Press. See also: Alder, J. and U.R. Sumaila. 2004. "Western Africa: A fish basket of Europe past and present." *Journal of Environment and Development* 13: 156–178.

[23] N.N. I wrote the foreword to a book that gives a fascinating account of the cod and turbot conflicts. The latter was largely engineered by an ambitious Canadian politician to blame Spain and to get elected as leader of a province, both of which he succeeded in doing (Pauly, D. 2003. "Foreword/Avant-propos." In: *Une taupe chez les morues: Halieuscopie d'un conflit*, by De Saint Pélissac, 5–9. Bleus Marines I. Mississauga, ON: AnthropoMare.)

[24] Ainley, D. and D. Pauly. 2013. "Fishing down the food web of the Antarctic continental shelf and slope." *Polar Record*, 50: 92-107.

[25] Pauly, D., J. Alder, A. Bakun, S. Heileman, K.H.S. Kock, P. Mace, W. Perrin, K.I. Stergiou, U.R. Sumaila, M. Vierros, K.M.F. Freire, Y. Sadovy, V. Christensen, K. Kaschner, M.L.D. Palomares, P. Tyedmers, C. Wabnitz, R. Watson, and B. Worm. 2005. "Marine Fisheries Systems." In: *Ecosystems and Human Well-being: Current States and Trends*, Vol. 1, edited by R. Hassan, R. Scholes, N. Ash, 477–511. Washington, DC: Millennium Ecosystem Assessment and Island Press.

[26] Swartz, W., E. Sala, R. Watson, and D. Pauly. 2010. "The spatial expansion and ecological footprint of fisheries (1950 to present)." PLOS ONE 5(12): e15143.

[27] N.N. Gelchu, A. and D. Pauly. 2007. "Growth and distribution of port-based fishing effort within countries' EEZs from 1970 to 1995." *Fisheries Centre Research Reports* 15(4). Vancouver, BC: University of British Columbia.

[28] Clarke, S.C., M.K. McAllister, E.J. Milner-Gulland, G.P. Kirkwood, C.G.J. Michielsens, D.J. Agnew, E.D. Pikitch, H. Nakano, and M.S. Shivji. 2006. "Global estimates of shark catches using trade records from commercial markets." *Ecological Letters* 9: 1115–1126. See also: Biery, L., and D. Pauly. 2012. "A global review of species-specific shark-fin-to-body-mass ratios and relevant legislation." *Journal of Fish Biology* 80: 1643–1677.

[29] Myers, R.A. and B. Worm. 2003. "Rapid worldwide depletion of predatory fish communities." *Nature* 423: 280–283.

[30] Floyd, J. and D. Pauly. 1984. "Smaller size tuna around the Philippines: Can fish aggregating devices be blamed?" *Infofish Marketing Digest* 5: 25-27.

[31] N.N. The increase of the depth of operations of fishing vessels was underestimated by Morato, T., R. Watson, T. Pitcher, and D. Pauly. 2006. "Fishing down the deep." *Fish and Fisheries* 7(1): 24–34. This was corrected in Watson, R.A. and T. Morato. 2013. "Fishing down the deep: Accounting for within-species changes in depth of fishing." *Fisheries Research* 140: 63–65. The latter method is incorporated in the derivation of Sea Around Us catch maps.

[32] Pauly, D., J. Alder, E. Bennett, V. Christensen, P. Tyedmers, and R. Watson. 2003. "The future for fisheries." *Science* 302: 1359–1361.

[33] Morato, T. R. Watson, T.J. Pitcher, and D. Pauly. 2006. "Fishing down the deep." *Fish and Fisheries* 7(1): 24-34.

[34] N.N. See also: Norse, E.A., S. Brooke, W.W.L. Cheung, M.R. Clark, I. Ekeland, R. Froese, K.M. Gjerde, R.L. Haedrich, S. S. Heppell, T. Morato, L.E. Morgan, D. Pauly, U.R. Sumaila, and R. Watson. 2012. "Sustainability of deep-sea fisheries." *Marine Policy* 36: 307-320.

[35] N.N. This makes the certification, in late 2016, of the New Zealand orange roughy fishery as "sustainable" by the Marine Stewardship Council particularly outrageous. Orange roughy (*Hoplostethus atlanticus*) fisheries, targeting a deep-water, long-lived fish, not only destroy habitats, but are heavily subsidized and generally resemble desperate mining operations; see, e.g.: Foley, N.S., T.M. van Rensburg, and C.W. Armstrong. 2011. "The rise and fall of the Irish orange roughy fishery: An economic analysis." *Marine Policy* 35: 756-763.

[36] Stergiou, K.I. 2002. "Overfishing, tropicalization of fish stocks, uncertainty and ecosystem management: Re-sharpening Ockham's razor." *Fisheries Research* 55: 1-9.

[37] Pauly, D. and R. Watson. 2005. "Background and interpretation of the 'Marine Trophic Index' as a measure of biodiversity." *Philosophical Transactions of the Royal Society B: Biological Sciences* 360: 415-423.

[38] N.N. The original version of this contribution had a caveat stating that the decline of MTI applied only to "fish above trophic levels of 3.5" (as shown in figure 2 of Branch et al. 2010, *Nature*, doi: 10.1038/nature09528), but this caveat turned out to be superfluous, as mean trophic levels, globally, do show a decline, even when low-trophic-level fish and invertebrates are included (http://www.seaaroundus.org and http://www.fishing.org).

[39] Pauly, D., V. Christensen, J. Dalsgaard, R. Froese, and F.C. Torres. 1998. "Fishing down marine food webs." *Science* 279: 860-863.

[40] Pauly, D., J. Alder, E. Bennett, V. Christensen, P. Tyedmers, and R. Watson. 2003. "The future for fisheries." *Science* 302: 1359-1361.

[41] Jacquet, J. and D. Pauly. 2008. "Trade secrets: Renaming and mislabeling of seafood." *Marine Policy* 32: 309-318.

[42] Jacquet, J. and D. Pauly. 2007. "The rise of seafood awareness campaigns in an era of collapsing fisheries." *Marine Policy* 31: 308-313.

[43] N.N. For example, the misleading of consumers by the Marine Stewardship Council, or MSC, is getting quite brazen, with the mining of the accumulated biomass of long-lived fishes such as orange roughy (*Hoplostethus atlanticus*) by New Zealand trawlers being certified as

"sustainable." Indeed, even the World Wide Fund for Nature (WWF), which helped create the MSC, now questions its very mode of operation and incentive structure, as evidenced in a draft report leaked in December 2016. See also previous note.

44 Bonfil, R., G. Munro, U.R. Sumaila, H. Valtysson, M. Wright, T.J.M. Pitcher, D. Preikshot, N. Haggan, and D. Pauly. 1998. "Impacts of distant water fleets: An ecological, economic and social assessment." In: *The Footprint of Distant Water Fleets on World Fisheries*, Endangered Seas Campaign, WWF International, 11-111. Godalming, Surrey: World Wide Fund for Nature.

45 N.N. This is well documented, for the United States, by Weber, M.L. 2002. *From Abundance to Scarcity: A History of US Marine Fisheries Policy*. Washington, DC: Island Press.

46 Kaczyinski, V.M. and D.L. Fluharty. 2002. "European policies in West Africa: Who benefits from fisheries agreements?" *Marine Policy* 26: 75–93.

47 N.N. See, e.g.: Belhabib, D., A. Mendy, Y. Subah, N. T Broh, A.S Jueseah, N. Nipey, N. Willemse, D. Zeller, and D. Pauly. 2016. "Fisheries catch under-reporting in The Gambia, Liberia and Namibia and the three Large Marine Ecosystems which they represent." *Environmental Development* 17: 157–174.

48 N.N. See: Belhabib D., U.R. Sumaila, V.W.Y. Lam., D. Zeller, P. Le Billon, E.A. Kane, and D. Pauly. 2015. "Euro vs. Yuan: Comparing European and Chinese fishing access in West Africa." PLOS ONE 10(3): e0118351.

49 N.N. See online supporting material of: Pauly, D., D. Belhabib, R. Blomeyer, W.W.L. Cheung, A. Cisneros-Montemayor, D. Copeland, S. Harper, V.W.Y. Lam, Y. Mai, F. Le Manach, H. Österblom, K.M. Mok, L. van der Meer, A. Sanz, S. Shon, U.R. Sumaila, W. Swartz, R. Watson, Y. Zhai, and D. Zeller. 2014. "China's distant-water fisheries in the 21st century." *Fish and Fisheries* 15: 474–488.

50 Komatsu, M. and S. Misaki. 2003. *Whales and the Japanese: How We Have Come to Live in Harmony with the Bounty of the Sea*. Tokyo: The Institute of Cetacean Research.

51 Chavance, P., M. Ba, D. Gascuel, M. Vakily, and D. Pauly, eds. 2004. *Marine Fisheries, Ecosystems and Society in West Africa: Half a Century of Change/ Pêcheries Maritimes, Écosystèmes et Sociétés en Afrique de l'Ouest: Un Demi-siècle de Changement*. Actes du symposium international, Dakar-Sénégal, 24–28 juin 2002. Office des publications officielles des communautés européennes, XXXVI, collection des rapports de recherche halieutique

ACP-UE 15. (All chapters, whether in French or English, have abstracts and figure and table captions in both languages.)

[52] Gerber, L., L. Morissette, K. Kaschner, and D. Pauly. 2009. "Should whales be culled to increase fishery yields?" *Science* 323: 880–881.

[53] Swartz, W. and D. Pauly. 2008. *Who's Eating All the Fish? The Food Security Rationale for Culling Cetaceans.* Washington, DC: Humane Society of the United States. (Available from: http://www.hsus.org/marine_mammals/save_whales_not_whaling/.)

[54] Mace, P.M. 1997. "Developing and sustaining world fisheries resources: The state of science and management." In: *Developing and Sustaining World Fisheries Resources: The State of Science and Management, Proceedings of Second World Fisheries Congress, Brisbane, Australia,* edited by D.H. Hancock, D.C. Smith, A. Grant, and J.P. Beumer, 1–20. Collingwood, Australia: CSIRO Publishing.

[55] Pauly, D., V. Christensen, S. Guénette, T.J. Pitcher, U.R. Sumaila, C.J. Walters, R. Watson, and D. Zeller. 2002. "Towards sustainability in world fisheries." *Nature* 418: 689–695.

[56] Anticamara, J.A., R. Watson, A. Gelchu, and D. Pauly. 2011. "Global fishing effort (1950–2010): Trends, gaps, and implications." *Fisheries Research* 107: 131–136.

[57] Pauly, D. and M.L.D. Palomares. 2010. "An empirical equation to predict annual increases in fishing efficiency." Fisheries Centre Working Paper #2010-07. (Available from: http://oceans.ubc.ca/publications/working-papers/.)

[58] Myers, R.A. and B. Worm. 2003. "Rapid worldwide depletion of predatory fish communities." *Nature* 423: 280–283.

[59] Christensen V., S. Guénette, J.J. Heymans, C.J. Walters, R. Watson, D. Zeller, and D. Pauly. 2003. "Hundred year decline of north Atlantic predatory fishes." *Fish and Fisheries* 4: 1–24.

[60] Roberts, C. 2007. *The Unnatural History of the Sea.* Washington, DC: Island Press.

[61] Rosenberg, A.A., W.J. Bolster, K.E. Alexander, W.B. Leavenworth, A.B. Cooper, and M.G. McKenzie. 2005. "The history of ocean resources: Modeling cod biomass using historical records." *Frontiers in Ecology and Evolution* 3(2): 84–90.

[62] Thurstan, R.H., S. Brockington, and C. Roberts. 2010. "The effects of 118 years of industrial fishing on UK bottom trawl fisheries." *Nature Communications* 1. doi:10.1038/ncomms1013.

[63] Sáenz-Arroyo, A., C.M. Roberts, J. Torre, M. Cariño-Olvera, and R. Enríquez-Andrade. 2005. "Rapidly shifting environmental baselines among fishers of the Gulf of California." *Philosophical Transactions of the Royal Society B: Biological Sciences* 272: 1957–1962.

[64] Alder, J., B. Campbell, V. Karpouzi, K. Kaschner, and D. Pauly. 2008. "Forage fish: From ecosystems to markets." *Annual Reviews of Environment and Resources* 33: 153–166; and Alder, J. and D. Pauly, eds. 2006. *On the Multiple Uses of Forage Fish: From Ecosystem to Markets.* Fisheries Centre Research Reports 14(3). Vancouver, BC: University of British Columbia.

[65] N.N. Cashion, T., F. Le Manach, D. Zeller, and D. Pauly. 2017. "Most fish destined for fishmeal production are food-grade fish." *Fish and Fisheries* 18. doi: 10.1111/faf.12209.

[66] Hites, R.A., J.A. Foran, D.O. Carpenter, M.C. Hamilton, B.A. Knuth, and S.J. Schwager. 2004. "Global assessment of organic contaminants in farmed salmon." *Science* 303: 225–229.

[67] N.N. Christensen, V., C. Piroddi, M. Coll, J. Steenbeek, J. Buszowski, and D. Pauly. 2014. "A century of fish biomass decline in the ocean." *Marine Ecology Progress Series* 512: 155–166.

[68] Stergiou, K.I., A.C. Tsikliras, and D. Pauly. 2009. "Farming up the Mediterranean food webs." *Conservation Biology* 23(1): 230–232.

[69] Bearzi, G.E., E. Politi, S. Agazzi, and A. Azzelino. 2006. "Prey depletion caused by overfishing and the decline of marine megafauna in eastern Ionian Sea coastal waters (central Mediterranean)." *Biological Conservation* 127: 373–382.

[70] FAO. 2010. *The State of World Aquaculture and Fisheries.* Rome: Food and Agriculture Organization of the United Nations.

[71] Kent, G. 2003. "Fish trade, food security, and the human right to adequate food." In: *Report of the Expert Consultation on International Fish Trade and Food Security.* FAO Fisheries Report 708, 49–70. Casablanca, Morocco, January 27–30, 2003. Rome: Food and Agriculture Organization of the United Nations.

[72] Alder, J. and U.R. Sumaila. 2004. "Western Africa: A fish basket of Europe past and present." *Journal of Environment and Development* 13: 156–178.

[73] Swartz, W., U.R. Sumaila, R. Watson, and D. Pauly. 2010. "Sourcing seafood from three major markets: The EU, Japan and the USA." *Marine Policy* 34(6): 1366–1373.

[74] Jacquet, J. and D. Pauly. 2007. "The rise of seafood awareness campaigns in an era of collapsing fisheries." *Marine Policy* 31: 308–313.

[75] Jacquet, J. and D. Pauly. 2008. "Trade secrets: Renaming and mislabeling of seafood." *Marine Policy* 32: 309–318.

[76] Jacquet, J., D. Pauly, D. Ainley, S. Holt, P. Dayton, and J.B.C. Jackson. 2010. "Seafood stewardship in crisis." *Nature* 467: 28–29.

[77] Jacquet, J. 2011. "Beyond food: Fish in the twenty-first century." In *Ecosystem Approaches to Fisheries: A Global Perspective*, edited by V. Christensen and J. Maclean, 120–127. Cambridge, UK: Cambridge University Press.

[78] Sumaila, U.R., A. Khan, A.J. Dyck, R. Watson, G. Munro, P. Tyedmers, and D. Pauly. 2010. "A bottom-up re-estimation of global fisheries subsidies." *Journal of Bioeconomics* 12: 201–225.

[79] N.N. The upper limit of the range, 35 billion USD, is a recent estimate, which does not differ much from the estimates documented in the contribution cited in the preceding note, once inflation is taken into account. See: Sumaila, U.R., V.W.I. Lam, F. Le Manach, W. Swartz, and D. Pauly. 2016. "Global fisheries subsidies: An updated estimate." *Marine Policy* 69: 189–193. doi: 1016/j.marpol.2015.12.026.

[80] Milazzo, M. 1998. *Subsidies in World Fisheries: A Re-examination*. World Bank Technical Paper 406. Fisheries Series. Washington, DC: The World Bank.

[81] Sumaila, U.R. and D. Pauly. 2007. "All fishing nations must unite to cut subsidies." *Nature* 450: 945.

[82] Pauly, D., J. Alder, E. Bennett, V. Christensen, P. Tyedmers, and R. Watson. 2003. "The future for fisheries." *Science* 302: 1359–1361.

[83] Watson, R. and D. Pauly. 2001. "Systematic distortions in world fisheries catch trends." *Nature* 414: 534–536.

[84] Pang, L. and D. Pauly. 2001. "Chinese marine capture fisheries from 1950 to the late 1990s: The hopes, the plans and the data." In: *The Marine Fisheries of China: Development and Reported Catches*, edited by R. Watson, L. Pang, and D. Pauly, 1–27. *Fisheries Centre Research Reports* 9(2). Vancouver, BC: University of British Columbia

[85] Batson, A. 2010. "Chinese data 'man-made.'" *Wall Street Journal*, World News: Asia, December 7.

[86] FAO. 2010. *The State of World Aquaculture and Fisheries*. Rome: Food and Agriculture Organization of the United Nations.

[87] N.N. See: Pauly, D. and D. Zeller. 2016. "Catch reconstructions reveal that global marine fisheries catches are higher than reported and declining." *Nature Communications* 7. doi: 10.1038/ncomms10244.

[88] N.N. This was masterfully explained in Oreskes N. and E.M. Conway. 2014. *The Collapse of Western Civilization: A View from the Future.* New York: Columbia University Press.

[89] Ludwig, D., R. Hilborn, and C.J. Walters. 1993. "Uncertainty, resource exploitation and conservation: Lessons from history." *Science* 260: 17 and 36.

[90] Friel, H. 2011. *The Lomborg Deception: Setting the Record Straight about Global Warming.* New Haven: Yale University Press.

[91] Oreskes, N. and E.M. Conway. 2010. *Merchants of Doubt: How a Handful of Scientists Obscured the Truth on Issues from Tobacco Smoke to Global Warming.* New York: Bloomsbury Press.

[92] Pauly, D. 2006. "Major trends in small-scale marine fisheries, with emphasis on developing countries, and some implications for the social sciences." *Maritime Studies* (MAST) 4(2): 7–22.

[93] N.N. See one-page summary accounts from "Albania" to "Yemen." In: Pauly, D. and D. Zeller, eds. 2016. *Global Atlas of Marine Fisheries: A Critical Appraisal of Catches and Ecosystem Impacts.* Washington, DC: Island Press.

[94] Jacquet, J.L., H. Fox, H. Motta, A. Ngusaru, and D. Zeller. 2010. "Few data, but many fish: Marine small-scale fisheries catches for Mozambique and Tanzania." *African Journal of Marine Science* 32: 197–206.

[95] Zeller, D., S. Booth, and D. Pauly. 2007. "Fisheries contribution to GDP: Underestimating small-scale fisheries in the Pacific." *Marine Resources Economics* 21: 355–374.

[96] The Sea Around Us is the research initiative at the University of British Columbia in Vancouver, Canada, of which I am the Principal Investigator. See: Pauly, D. 2007. "The *Sea Around Us* project: Documenting and communicating global fisheries impacts on marine ecosystems." *AMBIO: A Journal of the Human Environment* 34(4): 290–295.

[97] N.N. Pauly, D. and D. Zeller. 2016. "Catch reconstructions reveal that global marine fisheries catches are higher than reported and declining." *Nature Communications,* 7. doi: 10.1038/ncomms10244.

[98] N.N. See the essay titled "On Reconstructing Catch Time Series."

[99] Pikitch, E.K., C. Santora, E.A. Babcock, A. Bakun, R. Bonfil, D.O. Conover, P. Dayton, and P. Doukakis. 2004. "Ecosystem-based fishery management." *Science* 305: 346–347.

[100] Cury, P.M., Y.J. Shin, B. Planque, J.M. Durant, J.M. Fromentin, S. Kramer-Schadt, N.C. Stenseth, M. Travers, and V. Grimm. 2008. "Ecosystem

oceanography for global change in fisheries." *Trends in Ecology and Evolution* 23: 338–346.

[101] Pauly, D., J. Alder, E. Bennett, V. Christensen, P. Tyedmers, and R. Watson. 2003. "The future for fisheries." *Science* 302: 1359–1361.

[102] Cheung, W.W.L., V.W.Y. Lam, J.L. Sarmiento, K. Kearney, R. Watson, D. Zeller, and D. Pauly. 2010. "Large-scale redistribution of maximum fisheries catch potential in the global ocean under climate change." *Global Change Biology* 16: 24–35.

[103] Cheung, W.W.L., J. Dunne, J. Sarmiento, and D. Pauly. 2011. "Integrating eco-physiology and plankton dynamics into projected changes in maximum fisheries catch potential under climate change." *ICES Journal of Marine Science* 68: 1008–1018.

[104] Pauly, D. 2010. *Gasping Fish and Panting Squids: Oxygen, Temperature and the Growth of Water-Breathing Animals*. Excellence in Ecology. Book 22. Oldendorf/Luhe, Germany: International Ecology Institute.

[105] Pauly, D. 1995. "Anecdotes and the shifting baseline syndrome of fisheries." *Trends in Ecology and Evolution* 10(10): 430.

[106] Gladwell, M. 2000. *The Tipping Point: How Little Things Can Make a Big Difference*. New York: Little, Brown and Company.

[107] Pauly. D. 1998. "Beyond our original horizons: The tropicalization of Beverton and Holt." *Reviews in Fish Biology and Fisheries* 8(3): 307–334.

[108] N.N. The Chinese government now views the high sea as its "blue granary." (E.g., see: Hongzhou, Z. "China's growing appetite for fish and fishing disputes in the South China Sea." *All China Review*, November 8, 2016. http://www.allchinareview.com/chinas-growing-appetite-for-fish-and-fishing-disputes-in-the-south-china-sea/.)

[109] Jackson, J.B.C., M.X. Kirby, W.H. Berger, K.A. Bjorndal, L.W. Botsford, B.J. Bourque, R. Cooke, J.A. Estes, T.P. Hughes, S. Kidwell, C.B. Lange, H.S. Lenihan, J.M. Pandolfi, C.H. Peterson, R.S. Steneck, M.J. Tegner, and R.R. Warner. 2001. "Historical overfishing and the recent collapse of coastal ecosystems." *Science* 293: 629–638.

[110] Costello, C., S.D. Gaines, and J. Lynham. 2008. "Can catch shares prevent fisheries collapses?" *Science* 321: 1678–1681.

[111] Pauly, D. 2008. "Agreeing with Daniel Bromley." *Maritime Studies* (MAST) 6(2): 27–28.

[112] Macinko, S. and D.W. Bromley. 2002. *Who Owns America's Fisheries?* Washington, DC: Island Press.

[113] Macinko, S. and D.W. Bromley. 2004. "Property and fisheries for the twenty-first century: Seeking coherence from legal and economic doctrine." *Vermont Law Review* 28: 623–661.

[114] Pauly, D., W. Graham, S. Libralato, L. Morissette, and M.L.D. Palomares. 2009. "Jellyfish in ecosystems, online databases, and ecosystem models." *Hydrobiologia* 616(1): 67–85.

[115] N.N. A wonderful example of how a marine ecosystem can recover, via a marine reserve, from the devastating impact of a polluting industry (the canning of sardines) into a rich ecosystem providing a range of services to various industries (fishing, whale-watching, other forms of non-extractive tourism) is given in: Palumbi, S.R. and S. Sotka. 2011. *The Death and Life of Monterey Bay: A Story of Revival*. Washington, DC: Island Press.

[116] Wood, L, L. Fish, J. Laughren, and D. Pauly. 2008. "Assessing progress towards global marine protection targets: Shortfalls in information and action." *Oryx* 42(3): 340–351.

[117] N.N. Boonzaier, L. and D. Pauly. 2016. "Marine protection targets: An updated assessment of global progress." *Oryx* 50(1): 27–35. See also: http://www.pewtrusts.org/en/projects/global-ocean-legacy.

[118] N.N. Essentially, all countries but the U.S. are members of the Montreal-based Convention on Biological Diversity.

[119] N.N. It may be important to note here that marine reserves not only protect the sedentary fish living within their limits, but also are highly likely to lead to the emergence of more sedentary habits among a subset of highly migratory fishes. See: Mee, J.A., S. Otto, and D. Pauly. 2017. "Evolution of movement rate increases the effectiveness of large marine reserves for the conservation of pelagic fishes." *Evolutionary Applications* 10. doi: 10.1111/eva.12460.

[120] Margalef, R. 1968. *Perspectives in Ecological Theory*. Chicago: University of Chicago Press.

[121] While some readers clearly liked the title of this contribution (originally published by *The New Republic* on October 7, 2009, and reprinted here with permission), some colleagues have been turned off by its alarmist tone. Thus, I use this opportunity to clarify something that only scientists who write for nonscientific journals and magazines know from experience: the authors of articles do not choose their titles, editors do. Thus, you may love or hate the word "Aquacalypse," but I had nothing to do with it.

The then-editor of *The New Republic*, Franklin Foer, had solicited this contribution, possibly because shortly before, I had been interviewed by his brother Jonathan Safran Foer, who at the time was working on his book titled *Eating Animals*.

[122] N.N. This essay was written in 2009, in the year that the swindler Bernard L. Madoff was tried, which renewed interest in Ponzi-like pyramid schemes.

[123] N.N. See: Jacquet, J. and D. Pauly. 2008. "Trade secrets: Renaming and mislabeling of seafood." *Marine Policy* 32: 309–318.

[124] N.N. The scientific name of this fish is *Hoplostethus atlanticus*; see: http://www.fishbase.org.

[125] N.N. The scientific name of this fish is *Dissostichus eleginoides*; see: http://www.fishbase.org.

[126] N.N. The scientific name of this fish is *Macruronus novaezelandiae*; see: http://www.fishbase.org.

[127] N.N. See: Weber, M.L. 2002. *From Abundance to Scarcity: A History of US Marine Fisheries Policy*. Washington, DC: Island Press.

[128] N.N. Sumaila, U.R., V.W.Y. Lam, F. Le Manach, W. Schwartz, and D. Pauly. 2016. "Global fisheries subsidies: An updated estimate." *Marine Policy* 69: 189–193. doi: 1016/j.marpol.2015.12.026.

[129] N.N. The almost 90-million-ton peak in 1996 was the sum of nominal catches reported to FAO by its member countries; the actual catch in 1996 was higher, almost 140 million tons, and has also been declining since. See: http://www.seaaroundus.org and Pauly, D. and D. Zeller. 2016. "Catch reconstructions reveal that global marine fisheries catches are higher than reported and declining." *Nature Communications* 7. doi: 10.1038/ncomms10244.

[130] N.N. The study was that of Worm, B., E.B. Barbier, N. Beaumont, J.E. Duffy, C. Folke, B.S. Halpern, J.B.C. Jackson, H.K. Lotze, F. Micheli, S.R. Palumbi, E.E. Sala, K.A. Selkoe, J.J. Stachowicz, and R. Watson. 2006. "Impacts of biodiversity loss on ocean ecosystem services." *Science* 314: 787–790, and, although it garnered an extremely wide press coverage, most of it was misleading. This article projected that all fish exploited by commercial fisheries would be "collapsed" by the mid-21st century, meaning, in the context of that article, *that their populations would be yielding catches equal to or less than 10 percent of their historic maximum*, which is already the case for many, if not most of traditionally exploited populations (e.g., cod off New England). Yet, the press understood the species in question would go extinct, which is nonsense, and which allowed the detractors of the study to score points in the ensuing debate.

[131] Assume that 6 of the 12 fish populations that are being monitored in a given area are in "good shape." Now assume that 4 fish populations that have disappeared in the past were also counted: it would increase the denominator from 12 to 16, but not the numerator (because populations that have disappeared are not in good shape). The fraction of populations in good shape would then be 6/16, which is lower than 6/12.

[132] N.N. See the essay titled "The Shifting Baseline Syndrome of Fisheries" for an elaboration of this concept.

[133] N.N. This description certainly applies to me, although I was trained as a fisheries scientist (see the final three essays in this book).

[134] N.N. I do not recall, in late 2016, the World Bank document that I read in 2009 (perhaps it was the *Sunken Billions* study of 2009). However, I did find a more recent World Bank document whose summary says, among other things, that "[s]mall-scale fishing communities are among the poorest and most afflicted with social ills and may be further marginalized by a failure to recognize the importance of fisheries"; see: World Bank. 2012. *Hidden Harvest: The Global Contribution of Capture Fisheries.* Washington, DC: The World Bank.

[135] N.N. A rigorous documentation of the global increase of gelatinous zooplankton is provided in Brotz, L., W.W.L. Cheung, K. Kleisner, E. Pakhomov, and D. Pauly. 2012. "Increasing jellyfish populations: Trends in large marine ecosystems." *Hydrobiologia* 690(1): 3–20.

[136] N.N. See: Richardson, A.J., A. Bakun, G.C. Hays, and M.J. Gibbons. 2009. "The jellyfish joyride: Causes, consequences and management responses to a more gelatinous future." *Trends in Ecology & Evolution*, 24: 312–322.

[137] N.N. See, e.g.: Diaz, R.J. and R. Rosenberg. 2008. "Spreading dead zones and consequences for marine ecosystems." *Science* 15:926–929.

[138] N.N. Notably, individual fish that survive are expected to shrink (see: Cheung, W.W.L., J.L. Sarmiento, J. Dunne, T.L. Frölicher, V. Lam, M.L.D. Palomares, R. Watson, and D. Pauly. 2013. "Shrinking of fishes exacerbates impacts of global ocean changes on marine ecosystems." *Nature Climate Change* 3: 254–258) and fisheries to decline (see: Cheung, W.W.L., V.W.Y. Lam, J.L. Sarmiento, K. Kearney, R. Watson, D. Zeller, and D. Pauly. 2010. "Large-scale redistribution of maximum fisheries catch potential in the global ocean under climate change." *Global Change Biology* 16: 24–35).

[139] N.N. This is because marine fish have been shown to precipitate carbonates within their gut and then to excrete calcium carbonate, which tends to buffer acidification; hence high fish biomass would, at least in part,

mitigate ocean acidification; see: Wilson, R.W., F.J. Millero, J.R. Taylor, P.J. Walsh, V. Christensen, S. Jennings, and M. Grosell. 2009. "Contribution of fish to the marine inorganic carbon cycle." *Science* 323: 359–362.

[140] N.N. Lam, M. and D. Pauly. 2010. "Who is right to fish? Evolving a social contract for ethical fisheries." *Ecology and Society* 15: 16. http://www.ecologyandsociety.org/vol15/iss3/art16/.

[141] N.N. See: Cashion, T., F. Le Manach, D. Zeller, and D. Pauly. 2017. "Most fish destined for fishmeal production are food-grade fish." *Fish and Fisheries* 18. doi: 10.1111/faf.12209.

[142] N.N. Unfortunately, nothing has occurred in the years since this was originally written that would soften this assessment. Indeed, the MSC has become quite shameless; see: Jacquet, J., D. Pauly, D. Ainley, S. Holt, P. Dayton, and J. Jackson. 2010. "Seafood stewardship in crisis." *Nature* 467: 28–29.

[143] N.N. See: Jacquet, J. and D. Pauly. 2007. "The rise of seafood awareness campaigns in an era of collapsing fisheries. *Marine Policy* 31: 308–313.

[144] N.N. According to the UNCLOS, the vessels of foreign countries can fish in one's Exclusive Economic Zone, but this requires an explicit "access agreement" and the payment of an access fee.

[145] NN. This contribution, reprinted here with permission, was originally published as Pauly, D. 2006. "Major trends in small-scale marine fisheries, with emphasis on developing countries, and some implications for the social sciences." *Maritime Studies* (MAST) 4(2): 7–22. It was based on the keynote address given on July 7, 2005, at the conference on "People and the Seas III: New Directions in Coastal and Maritime Studies," held at the Centre for Maritime Research of the University of Amsterdam, and acknowledged the many colleagues and friends from the social sciences, notably the late Bob Johannes and Ken Ruddle, who introduced me to subtleties not accessible through the quantitative models of fisheries "stock assessments," and Derek Johnson, for inviting this contribution. I also thanked Deng Palomares for her help in the preparation of this paper, and Dirk Zeller, Jackie Alder, Rashid Sumaila, and two anonymous reviewers for critical comments.

[146] Pauly, D., V. Christensen, S. Guénette, T.J. Pitcher, U.R. Sumaila, C.J. Walters, R. Watson, and D. Zeller. 2002. "Towards sustainability in world fisheries." *Nature* 418: 689–695.

[147] Mace, P.M. 1997. "Developing and sustaining world fisheries resources: The state of science and management." In: *Developing and Sustaining World Fisheries Resources: The State of Science and Management*, edited by

D.H. Hancock, D.C. Smith, A. Grant, and J.P. Beumer, 1–20. Proceedings of Second World Fisheries Congress, Brisbane, Australia. Collingwood, Australia: CSIRO Publishing.

[148] Thompson, D. and FAO. 1988. "The world's two marine fishing industries—How they compare." *Naga, the* ICLARM *Quarterly* 11(3): 17.

[149] Pauly, D. 1997. "Small-scale fisheries in the tropics: Marginality, marginalization and some implications for fisheries management." In: *Global Trends: Fisheries Management*, edited by E.K. Pikitch, D.D. Huppert, and M.P. Sissenwine, 40–49. Proceedings from American Fisheries Society Symposium 20. Bethesda, MD: American Fisheries Society.

[150] Allison, E.H. and F. Ellis. 2001. "The livelihood approach and management of small-scale fisheries." *Marine Policy* 25: 377–388.

[151] Béné, C. 2003. "When fishery rhymes with poverty: A first step beyond the old paradigm on poverty in small-scale fisheries." *World Development* 31: 949–975.

[152] Pauly, D. 1997. "Small-scale fisheries in the tropics: Marginality, marginalization and some implications for fisheries management." In: *Global Trends: Fisheries Management*, edited by E.K. Pikitch, D.D. Huppert, and M.P. Sissenwine, 40–49. Proceedings from American Fisheries Society Symposium 20. Bethesda, MD: American Fisheries Society.

[153] Longhurst, A.R. and D. Pauly. 1987. *Ecology of Tropical Oceans*. San Diego: Academic Press.

[154] N.N. The Google Scholar records reported here were obtained shortly before the People and the Sea III conference was held (i.e., in June 2005). I redid the counts in late December 2016. The key result was the same: although the total number of records had increased by one order of magnitude, similar ratios were found for fisheries+ecology and fisheries+economics vs. fisheries+anthropology and fisheries+sociology.

[155] Pauly, D. 1994. *On the Sex of Fish and the Gender of Scientists: Essays in Fisheries Science*. London: Chapman & Hall.

[156] Jackson, J.B.C., M.X. Kirby, W.H. Berger, K.A. Bjorndal, L.W. Botsford, B.J. Bourque, R. Cooke, J.A. Estes, T.P. Hughes, S. Kidwell, C.B. Lange, H.S. Lenihan, J.M. Pandolfi, C.H. Peterson, R.S. Steneck, M.J. Tegner, and R.R. Warner. 2001. "Historical overfishing and the recent collapse of coastal ecosystems." *Science* 293: 629–638.

[157] Colonial Office. 1961. *Colonial Research, 1960–1961*. London: Her Majesty's Stationery Office.

[158] Butcher, J.G. 2004. *The Closing of the Frontier: A History of the Marine Fisheries of Southeast Asia c. 1850–2000*. Singapore: Institute of Southeast Asian Studies.

[159] Pauly, D. and J. Maclean. 2003. *In a Perfect Ocean: The State of Fisheries and Ecosystems in the North Atlantic Ocean*. Washington, DC: Island Press.

[160] Grainger, R.J.R. and S. Garcia. 1996. *Chronicles of Marine Fishery Landings (1950–1994): Trend Analysis and Fisheries Potential*. FAO Fisheries Technical Paper 359. Rome: Food and Agriculture Organization of the United Nations.

[161] Pauly, D., J. Alder, A. Bakun, S. Heileman, K.H.S. Kock, P. Mace, W. Perrin, K.I. Stergiou, U.R. Sumaila, M. Vierros, K.M.F. Freire, Y. Sadovy, V. Christensen, K. Kaschner, M.L.D. Palomares, P. Tyedmers, C. Wabnitz, R. Watson, and B. Worm. 2005. "Marine Fisheries Systems." In: *Ecosystems and Human Well-being: Current States and Trends*, Vol. 1., edited by R. Hassan, R. Scholes, and N. Ash, 477–511. Washington, DC: Millennium Ecosystem Assessment and Island Press.

[162] Bonfil, R., G. Munro, U.R. Sumaila, H. Valtysson, M. Wright, T.J.M. Pitcher, D. Preikshot, N. Haggan, and D. Pauly. 1998. "Impacts of distant water fleets: An ecological, economic and social assessment." In: *The Footprint of Distant Water Fleets on World Fisheries*, Endangered Seas Campaign, WWF International, II–III. Godalming, Surrey: WWF International.

[163] Alder, J. and U.R. Sumaila. 2004. "Western Africa: A fish basket of Europe past and present." *Journal of Environment & Development* 13: 156–178.

[164] N.N. East-West, in the period alluded to, meant the Soviet Union and its allies, notably in Eastern Europe vs. the U.S. and its allies, notably in Western Europe; China was isolated, if nominally part of the "Eastern Bloc."

[165] Firth, R. 1946. *Malay Fishermen: Their Peasant Economy*. London: Keagan.

[166] Ruddle, K. and R.E. Johannes, eds. 1985. *The Traditional Knowledge and Management of Coastal Systems in Asia and the Pacific*. Jakarta: UNESCO.

[167] Dyer, C.L. and R.M. McGoodwin, eds. 1994. *Folk Management in the World's Fisheries: Lessons for Modern Fisheries Management*. Niwot: University Press of Colorado.

[168] Butcher, J.G. 2004. *The Closing of the Frontier: A History of the Marine Fisheries of Southeast Asia c. 1850–2000*. Singapore: Institute of Southeast Asian Studies.

[169] Pauly, D. 1996. "Biodiversity and the retrospective analysis of demersal trawl surveys: A programmatic approach." In: *Baseline Studies in Biodiversity: The Fish Resources of Western Indonesia*, edited by D. Pauly and P. Martosubroto, 1–6. ICLARM Studies and Reviews 23. Manila: International Center for Living Aquatic Resources Management.

[170] Panayotou, T. and S. Jetanavarich. 1987. *The Economics and Management of Thai Marine Fisheries.* ICLARM Studies and Reviews 14. Manila: International Center for Living Aquatic Resources Management.

[171] Butcher, J.G. 2002. "Getting into trouble: The diaspora of Thai trawlers, 1975–2002." *International Journal of Maritime History* 14(2): 85–121.

[172] Lawson, R. and E. Kwei. 1974. *African Entrepreneurship and Economic Growth: A Case Study of the Fishing Industry of Ghana.* Accra: Ghana Universities Press.

[173] Atta-Mills, J., J. Alder, and U.R. Sumaila. 2004. "The decline of a regional fishing nation: The case of Ghana in West Africa." *Natural Resources Forum* 28: 13–21.

[174] Agüero, M., ed. 1992. *Contribuciones para el Studio de la Pesca Artesanal en América Latina.* ICLARM Conference Proceedings 35. Manila: International Center for Living Aquatic Resources Management.

[175] Hersoug, B. 2004. "Exporting fish, importing institutions: Fisheries development in the Third World." In *Fisheries Development: The Institutional Challenge,* edited by B. Hersoug, S. Jentoft, and P. Degnbol, 21–92. Delft: Eburon.

[176] Alverson, D.L., M. Freeberg, S.A. Murawski, and J.G. Pope. 1994. *A Global Assessment of Fisheries Bycatch and Discards.* FAO Fisheries Technical Paper 339. Rome: Food and Agriculture Organization of the United Nations.

[177] Kelleher, K. 2005. *Discards in the World's Marine Fisheries: An Update.* FAO Fisheries Technical Paper 470. Rome: Food and Agriculture Organization of the United Nations.

[178] Zeller, D. and D. Pauly. 2005. "Good news, bad news: Global fisheries discards are declining, but so are total catches." *Fish and Fisheries* 6: 156–159.

[179] Pauly, D., V. Christensen, S. Guénette, T.J. Pitcher, U.R. Sumaila, C.J. Walters, R. Watson, and D. Zeller. 2002. "Towards sustainability in world fisheries." *Nature* 418: 689–695.

[180] Zeller, D., S. Booth, P. Craig, and D. Pauly. 2006. "Reconstruction of coral reef fisheries catches in American Samoa, 1950–2002." *Coral Reefs* 25: 144–152.

[181] Chuenpagdee, R. and D. Pauly. 2006. "Small is beautiful? A database approach for global assessment of small-scale fisheries: Preliminary results and hypotheses." In: *Proceedings of the 4th World Fisheries Congress,* May 2004, Vancouver, Canada, edited by J. Nielsen, J.J. Dodson, K. Friedland, T.R. Hamon, J. Musick, and E. Verspoor, 587–595. Bethesda, MD: American Fisheries Society.

[182] Anonymous. 1997. *The Pacific's Tuna: The Challenge of Investing in Growth.* Manila: Office of Pacific Operations, Asian Development Bank.

[183] Gillett, R., L. McCoy, J. Rodwell, and J. Tamate. 2001. *Tuna: A Key Economic Resource in the Pacific.* Pacific Studies Series. Manila: Asian Development Bank, Forum Fisheries Agency.

[184] Dalzell, P., T.J.H. Adams, and N.V.C. Polunin. 1996. "Coastal fisheries in the Pacific Islands." *Oceanographic Marine Biology Annual Review* 34: 395–531.

[185] Chapman, M.D. 1987. "Women fishing in Oceana." *Human Ecology* 15: 267–288.

[186] Dalzell, P., T.J.H. Adams, and N.V.C. Polunin. 1996. "Coastal fisheries in the Pacific Islands." *Oceanographic Marine Biology Annual Review* 34: 395–531.

[187] Johannes, R.E. 1981. *Words of the Lagoon: Fishing and Marine Lore in the Palau District of Micronesia.* Berkeley: University of California Press.

[188] Gillett, R. and C. Lightfoot. 2002. *The Contribution of Fisheries to the Economies of Pacific Island Countries.* Pacific Study Series. Manila: Asian Development Bank.

[189] Zeller, D., S. Booth, P. Craig, and D. Pauly. 2006. "Reconstruction of coral reef fisheries catches in American Samoa, 1950–2002." *Coral Reefs* 25: 144–152.

[190] N.N. See: Zeller, D., S. Harper, K. Zylich, and D. Pauly. 2015. "Synthesis of under-reported small-scale fisheries catch in Pacific-island waters." *Coral Reefs* 34(1): 25–39.

[191] N.N. Just for information: The word "Spam" is derived from "spicy ham," canned meat that was part of the food supply of U.S. soldiers in WWII and that became very popular in the Pacific Islands.

[192] Pauly, D. 1997. "Small-scale fisheries in the tropics: Marginality, marginalization and some implications for fisheries management." In: *Global Trends: Fisheries Management,* edited by E.K. Pikitch, D.D. Huppert, and M.P. Sissenwine, 40–49. American Fisheries Society Symposium 20. Bethesda, MD: American Fisheries Society.

[193] Pauly, D. 2005. "Rebuilding fisheries will add to Asia's problems (Correspondence)." *Nature* 433: 457.

[194] Pearson, H. 2005. "Scientists seek action to fix Asia's ravaged ecosystems." *Nature* 433: 94.

[195] Longhurst, A.R. and D. Pauly. 1987. *Ecology of Tropical Oceans.* San Diego: Academic Press.

[196] Sarjono, I. 1980. "Trawlers banned in Indonesia." ICLARM Newsletter 3(4): 3.

[197] Pauly, D. 1997. "Small-scale fisheries in the tropics: Marginality, marginalization and some implications for fisheries management." In: Global Trends: Fisheries Management, edited by E.K. Pikitch, D.D. Huppert, and M.P. Sissenwine, 40-49. American Fisheries Society Symposium 20. Bethesda, MD: American Fisheries Society.

[198] Bailey, C. 1982. Small-Scale Fisheries of San Miguel Bay, Philippines: Occupational and Geographic Mobility. ICLARM Technical Report 10. Manila: International Center for Living Aquatic Resources Management.

[199] Baldauf, S. 2005. "Boat-building boom threatens Aceh fisheries." Christian Science Monitor, January 11. (Available from: https://www.csmonitor.com.)

[200] Chuenpagdee, R. 2005. "Business as usual for tsunami-affected communities in Thailand." Sea Around Us Newsletter (30): 1-3. (Available from: http://www.seaaroundus.org.)

[201] Erdmann, M. 2005. "Rebuilding Aceh's fishing fleets: Anecdotal field observations of an ill-conceived concept gone predictably astray." Sea Around Us Newsletter (31): 6. (Available from: http://www.seaaroundus.org.)

[202] Smith, I.R. 1981. "Improving fishing incomes when resources are overfished." Marine Policy 5(1): 17-22.

[203] Pauly, D., V. Christensen, S. Guénette, T.J. Pitcher, U.R. Sumaila, C.J. Walters, R. Watson, and D. Zeller. 2002. "Towards sustainability in world fisheries." Nature 418: 689-695.

[204] Pauly, D. 1997. "Small-scale fisheries in the tropics: Marginality, marginalization and some implications for fisheries management." In: Global Trends: Fisheries Management, edited by E.K. Pikitch, D.D. Huppert, and M.P. Sissenwine, 40-49. American Fisheries Society Symposium 20. Bethesda, MD: American Fisheries Society.

[205] This 1997 contribution, as of January 2017, has been cited 210 times, but overwhelmingly by biologists.

[206] Kaczynski, V.M. 2005. Presentation at the Liu Centre for Global Issues, University of British Columbia, Vancouver, May 2005.

[207] McManus, J.W., C.L. Nañola Jr., R.B. Reyes, and K.N. Kesner. 1992. "Resource Ecology of the Bolinao Coral Reef System." ICLARM Studies and Reviews 22. Manila: International Center for Living Aquatic Resources Management.

[208] Pauly, D. 1988. "Some definitions of overfishing relevant to coastal zone management in Southeast Asia." Tropical Coastal Area Management 3(1): 14-15.

209 N.N. Zeller, D., S. Booth, and D. Pauly. 2007. "Fisheries contributions to gross domestic products: Underestimating small-scale fisheries in the Pacific." *Marine Resources Economics* 21: 355–374.

210 Geertz, C. 1985. *Local Knowledge: Further Essays in Interpretive Anthropology.* New York: Basic Books.

211 Chapman, M.D. 1987. "Women fishing in Oceana." *Human Ecology* 15: 267–288.

212 Christensen V., S. Guénette, J.J. Heymans, C.J. Walters, R. Watson, D. Zeller, and D. Pauly. 2003. "Hundred year decline of north Atlantic predatory fishes." *Fish and Fisheries* 4: 1–24.

213 Pauly, D. and R. Watson. 2005. "Background and interpretation of the 'Marine Trophic Index' as a measure of biodiversity." *Philosophical Transactions of the Royal Society B: Biological Sciences* 360: 415–423.

214 Pauly, D., J. Alder, E. Bennett, V. Christensen, P. Tyedmers, and R. Watson. 2003. "The future for fisheries." *Science* 302: 1359–1361.

215 N.N. See: Pauly, D. 2018. "A vision of marine fisheries in a global blue economy." *Marine Policy* 87: 371–374.

216 N.N. I reprint here, with permission, a brief comment (Pauly, D. and A. Charles. 2015. "Counting on small-scale fisheries." *Science* 347: 242–243) in which one major constraint to policy making pertinent to artisanal fisheries was brought to the attention of the FAO and other relevant bodies:

"On 10 June 2014, the member States of the Food and Agriculture Organization of the United Nations (FAO) adopted the *Voluntary Guidelines for Securing Sustainable Small-Scale Fisheries in the Context of Food Security and Poverty Eradication* (1). To make these Guidelines effective, it is crucial that the FAO, governments, and civil society have access to data to help understand small-scale fisheries. Currently, catches from these fisheries are not collected separately, but are lumped in with industrial catches, even though they represent about one-quarter of global catches, and the majority of catches in many developing countries. To promote the transparency needed for good governance (2, 3), the FAO ought to request from member countries a report of catch data that distinguishes between industrial and small-scale fisheries. Many decades of debate have failed to produce one, agreed-upon definition of a 'small-scale fishery', but the modest variations in definitions between countries do not preclude efforts to gather global statistics. Just as the Guidelines do not impose a single definition of small-scale fisheries, each of the FAO's member States could define their own small-scale fisheries, reflecting local realities. These changes would help to highlight the importance of small-scale fisheries and may also help governments that still treat these fisheries as a solution to demographic pressure and rural landlessness (4) to focus instead on their inherent value."

References for the quoted passage:

1. FAO. 2014. "Voluntary Guidelines for Securing Sustainable Small-scale Fisheries in the Context of Food Security and Poverty Eradication." Rome: Food and Agriculture Organization of the United Nations.

2. Charles, A.T. 2011. "Small-scale fisheries: On rights, trade and subsidies." *Maritime Studies* (MAST) 10: 85–94.

3. Charles, A.T. 2013. "Governance of tenure in small-scale fisheries: Key considerations." *Land Tenure Journal* 1: 9–37.

4. Pauly, D. 2006. "Major trends in small-scale marine fisheries, with emphasis on developing countries, and some implications for the social sciences." *Maritime Studies* (MAST) 4: 7–22.

This comment was also communicated to a high-ranking official at FAO, who gracefully acknowledged its receipt and indicated that the issue it raised would be examined. We heard later that the official in question had retired, and moreover, that FAO, as a "technical organization" of the United Nations, is not permitted to collect statistics about entities whose definitions vary between countries. Thus, it is likely that the Sea Around Us catch database, presented here in the essay "A Global Community-Driven Catch Database," which relies on (not very) variable national definitions, will remain the only source of global marine small-scale fishery catch data.

[217] N.N. This contribution, reprinted here with permission, was originally published with the same title in *Reviews in Fish Biology and Fisheries* (1996; 6: 109–112). It was written to comment on four other contributions dealing with individual transferable quotas in the same journal issue. The articles' references are given in the notes below. It acknowledged Trevor Hutton for reading the draft and for suggestions that much improved it.

[218] Grafton, R.Q. 1996. "Individual transferable quotas: Theory and practice." *Reviews in Fish Biology and Fisheries* 6: 5–20.

[219] N.N. Nowadays, it requires no fantasy at all, as it has happened; see, e.g.: Pinkerton, E. and D.N. Edwards. 2009. "The elephant in the room: The hidden costs of leasing individual transferable fishing quotas." *Marine Policy*, 33(4): 707–713.

[220] Grafton, R.Q. 1996. "Individual transferable quotas: Theory and practice." *Reviews in Fish Biology and Fisheries* 6: 5–20.

[221] Arnason, R. 1996. "On the ITQ fisheries management system of Iceland." *Reviews in Fish Biology and Fisheries* 6: 63–90.

[222] N.N. Ragnar Arnason was wrong in this, as he was in many other things: the Icelandic ITQs were quickly concentrated in a few bankers' hands, then, as expected, they disappeared in the financial meltdown of 2008; see: Benediktsson K. and A. Karlsdóttir. 2011. "Iceland crisis and regional development: Thanks for all the fish?" *European Urban and Regional Studies* 18(2): 228–235.

[223] Annala, J.H. 1996. "New Zealand's ITQ system: Have the first eight years been a success or a failure?" *Reviews in Fish Biology and Fisheries* 6: 43–62.

[224] Walters, C.J. and R. H. Pearse. 1996. "Stock information requirements for quota management systems in commercial fisheries." *Reviews in Fish Biology and Fisheries* 6: 21–42.

[225] N.N. I can't resist mentioning that high-grading has recently been demonstrated to be as rampant in the ITQ fisheries of New Zealand as human rights abuses. (For the latter, see: Simmons, G. and C. Stringer 2014. "New Zealand's fisheries management system: Forced labour an ignored or overlooked dimension?" *Marine Policy* 50: 74–80.) The uncovering of massive high-grading (in: Simmons, G., G. Bremner, C. Stringer, B. Torkington, L.C.L. Teh, K. Zylich, D. Zeller, D. Pauly, and H. Whittaker. 2015. "Reconstruction of marine fisheries catches for New Zealand [1950–2010]." Fisheries Centre Working Paper #2015-87, Vancouver, BC: University of British Columbia) led to a whole series of other management failures (and a few environmental crimes) being uncovered (see, e.g.: Pala, C. 2017. "New Zealand fishing industry accused of driving world's rarest dolphin to extinction." *Earth Island Journal* Available from: http://www.earthisland.org/journal/index.php/elist/eListRead/new_zealand_fishing_industry_driving_maui_hectors_dolphin_to_extinction/).

[226] Warren, B. 1995. "Bycatch strategies: Success stories, promising approaches and role of the third sector." In: *Bycatches in Fisheries and Their Impact on the Ecosystem*, edited by T.J. Pitcher and R. Chuenpagdee. Fisheries Centre Research Reports 2(1): 61–64. Vancouver, BC: University of British Columbia.

[227] Arnason, R. 1996. "On the ITQ fisheries management system of Iceland." *Reviews in Fish Biology and Fisheries* 6: 63–90.

[228] Ibid.

[229] Christy, F.T., Jr. 1982. *Territorial Use Rights in Fisheries: Definitions and Conditions*. FAO Fisheries Technical Paper 227. Rome: Food and Agriculture Organization of the United Nations.

[230] N.N. This contribution, reprinted here with permission of the publisher, is based on a book chapter: Pauly, D. 1999. "Fisheries management:

Putting our future in places." In: *Fishing Places, Fishing People: Traditions and Issues in Canadian Small-Scale Fisheries*, edited by D. Newell and R. Ommer, 355-362. Toronto: University of Toronto Press. It acknowledged Dr. Anthony Davis for detailed and helpful comments on the structure and draft of that chapter.

[231] Annala, J. 1996. "New Zealand's ITQ system: Have the first eight years been a success or a failure?" *Reviews in Fishery Biology and Fisheries* 6: 43–62.

[232] Munro, G.R. and T.J. Pitcher, eds. 1996. "Individual transferable quotas." Special issue. *Reviews in Fishery Biology and Fisheries* 6: 1–116.

[233] N.N. Not to mention their inherent unfairness; see: Bromley, D.W. 2009. "Abdicating responsibility: The deceits of fisheries policy." *Fisheries* 34(6) 280–290.

[234] Pauly, D. 1980. "On the interrelationships between natural mortality, growth parameters and mean environmental temperature in 175 fish stocks." *Journal du Conseil International pour l'Exploration de la Mer* 39: 175–192.

[235] Walters, C. and P.H. Pearse. 1996. "Stock information requirements for quota management systems in commercial fisheries." *Reviews in Fishery Biology and Fisheries* 6: 21–42.

[236] Bohnsack, J.A., subcommittee chair. 1990. "The potential of marine fisheries reserves for reef fish management in the U.S. southern Atlantic." Snapper-Grouper Development Team Report to the South Atlantic Fishery Management Council. NOAA *Technical Memorandum* NMFS-SEFC-261.

[237] Hutchings, J.A. and R.A. Myers. 1994. "What can be learned from the collapse of a renewable resource? Atlantic cod, *Gadus morhua*, for Newfoundland and Labrador." *Canadian Journal of Fisheries and Aquatic Science* 51: 2126–2146.

[238] Myers, R.A., N.J. Barrowman, and K.R. Thompson. 1995. "Synchrony of recruitment across the North Atlantic: An update. (Or, 'now you see it, now you don't!')." *ICES Journal of Marine Science* 52: 103–110.

[239] This is because of the population growth of fish, which is suppressed when a fish population is very abundant, but increases when it is moderately fished, as first documented by Baranov, F.I. 1927; translation: F.I. Baranov. 1977. "More about the poor catch of roach." In: *Selected Works on Fishing Gears: Theory of Fishing*, Vol. 3, 62–64. Jerusalem: Israel Program for Scientific Translations.

[240] See, for a case study: Pauly, D. 1988. "Fisheries research and the demersal fisheries of Southeast Asia." In: *Fish Population Dynamics*, 2nd ed.,

edited by J.A. Gulland, 329–348. Chichester and New York: Wiley InterScience.

[241] See, for an attempt at a consensus statement: Roberts, C., W.J. Ballantine, C.D. Buxton, P. Dayton, L.B. Crowder, W. Milon, M.K. Orbach, D. Pauly, and J. Trexler. 1995. "Review of the use of marine fishery reserves in the U.S. Southeastern Atlantic." NOAA *Technical Memorandum* NMFS-SEFSC-376.

[242] See pages 365–368 of: Beverton, R.J.H. and S.J. Holt. 1957. *On the Dynamics of Exploited Fish Populations*. Fisheries Investigations, Series II. London: Ministry of Agriculture, Fisheries, and Food.

[243] Pauly, D. 1993. "Foreword." In Beverton, R.J.H. and S.J. Holt. *On the Dynamics of Exploited Fish Populations*. Reprint of the 1957 edition. London: Chapman & Hall.

[244] N.N. This essay, reprinted here with permission of the publisher, was originally published as Pauly, D. 1999. "Fisheries management: Putting our future in places." In: *Fishing Places, Fishing People: Traditions and Issues in Canadian Small-Scale Fisheries*, edited by D. Newell and R. Ommer, 355–362. Toronto: University of Toronto Press. It acknowledged Dr. Anthony Davis for detailed and helpful comments on the structure and content of the draft.

[245] Pauly, D. 1997. "Small scale fisheries in the tropics: Marginality, marginalization and some implications for fisheries management." In: *Global Trends in Fisheries Management*, edited by E. Pikitch, D.D. Hubert, and M. Sissenwine, 40–49. American Fisheries Society Symposium 20. Bethesda, MD: American Fisheries Society.

[246] Charles, A.T. 1995. "The Atlantic Canadian ground fishery: Roots of a collapse. *Dalhousie Law Journal* 18(1): 65–83.

[247] McGuire. T. 1991. "Science and the destruction of a shrimp fleet." *Maritime Anthropological Studies* 4(1): 32–55.

[248] Hutchings, J.A. and R.A. Myers. 1994. "What can be learned from the collapse of a renewable resource? Atlantic cod, *Gadus morhua*, for Newfoundland and Labrador." *Canadian Journal of Fisheries and Aquatic Sciences* 51: 2126–2146.

[249] Alverson, D.L., M.H. Freeberg, S.A. Murawski, and J.G. Pope. 1994. *A Global Assessment of Fisheries Bycatch and Discards*. FAO Fisheries Technical Paper 339. Rome: Food and Agriculture Organization of the United Nations.

[250] Finlayson, A.C. 1994. *Fishing for Truth: A Sociological Analysis of Northern Cod Assessments for 1977 to 1990*. St. John's, NL: The Institute of Social and Economic Research.

[251] Walters. C. 1994. *Fish on the Line: The Future of Canada's Pacific Fisheries.* Vancouver, British Columbia: David Suzuki Foundation.

[252] Pauly, D. 1995. "Anecdotes and the shifting baseline syndrome of fisheries." *Trends in Ecology and Evolution* 10(10): 430.

[253] Roy, D.J., B.E. Wynne, and R.W. Old, eds. 1991. *Bioscience ⇌ Society.* Chichester: John Wiley and Sons.

[254] Clark, C.W. 1990. *Mathematical Bioeconomics: The Optimal Management of Renewable Resources.* 2nd ed. New York: Wiley InterScience.

[255] Garcia, S. and C. Newton. 1997. "Current situation, trends, and prospects in world capture fisheries." In: *Global Trends in Fisheries Management,* edited by E. Pikitch, D.D. Hubert, and M. Sissenwine, 3–27. American Fisheries Society Symposium 20. Bethesda, MD: American Fisheries Society.

[256] Pauly, D. 1997. "Small scale fisheries in the tropics: Marginality, marginalization and some implications for fisheries management." In: *Global Trends in Fisheries Management,* edited by E. Pikitch, D.D. Hubert, and M. Sissenwine, 40–49. American Fisheries Society Symposium 20. Bethesda, MD: American Fisheries Society.

[257] Davis, A. 1996. "Barbed wire and bandwagons: A comment on ITQ fisheries management." *Reviews in Fish Biology and Fisheries* 6(1): 97–197.

[258] Amason, R. 1996. "On the ITQ fisheries management system in Iceland." *Reviews in Fish Biology and Fisheries* 6(1): 63–90.

[259] McCay, B.J. and J.M. Acheson, eds. 1987. *The Question of the Commons: The Culture and Ecology of Communal Resources.* Tucson: University of Arizona Press.

[260] Ruddle, K. and R.E. Johannes, eds. 1985. *The Traditional Knowledge and Management of Coastal Systems in Asia and the Pacific.* Jakarta: UNESCO Regional Office for Science and Technology.

[261] Pauly, D. 1988. "Fisheries research and the demersal fisheries of Southeast Asia." In: *Fish Population Dynamics,* 2nd ed., edited by J.A. Gulland, 329–348. Chichester and New York: Wiley InterScience.

[262] Finlayson, A.C. 1994. *Fishing for Truth: A Sociological Analysis of Northern Cod Assessments for 1977 to 1990.* St. John's, NL: The Institute of Social and Economic Research.

[263] Charles, A.T. 1995. "The Atlantic Canadian ground fishery: Roots of a collapse." *Dalhousie Law Journal* 18(1): 65–83.

[264] Pinkerton, E., ed. 1989. *Co-operative Management of Local Fisheries: New Directions for Improved Management and Community Development.* Vancouver, BC: UBC Press.

[265] Pinkerton, E. and M. Weinstein, eds. 1995. *Fisheries That Work: Sustainability through Community-Based Management*. Vancouver, BC: David Suzuki Foundation.

[266] Neis, B. 1999. "Familial and social patriarchy in the Newfoundland fishing industry." In: *Fishing Places, Fishing People: Traditions and Issues in Canadian Small-Scale Fisheries*, edited by D. Newell and R. Ommer, 32–54. Toronto, ON: University of Toronto Press.

[267] Kooiman, J., ed. 1992. *Modern Governance: New Government–Society Interactions*. London: Sage Publications.

[268] Baranov, F.I. 1977. *Selected Works on Fishing Gears: Theory of Fishing*, Vol. 3. Jerusalem: Israel Program for Scientific Translations.

[269] Beverton, R.J.H. and S.J. Holt. 1957. *On the Dynamics of Exploited Fish Populations*. Fisheries Investigations. Series II. London: Ministry of Agriculture, Fisheries, and Food.

[270] Cadigan, S.T. 1999. "Failed proposals for fisheries management and conservation in Newfoundland, 1855–1880." In: *Fishing Places, Fishing People: Traditions and Issues in Canadian Small-Scale Fisheries*, edited by D. Newell and R. Ommer, 147–169. Toronto, ON: University of Toronto Press.

[271] Bohnsack. 1994. "Marine reserves: They enhance fisheries, reduce conflicts, and protect resources." *Naga, the ICLARM Quarterly* 17(3): 4–7.

[272] Roberts. C., W.J. Ballantine, C.D. Buxton, P. Dayton, L.B. Crowder, W. Milon, M.K. Orbach, D. Pauly, and J. Trexler. 1995. "Review of the use of marine fishery reserves in the U.S. Southeastern Atlantic." *NOAA Technical Memorandum*, NMFS-SEFSC-376.

[273] Russ, G. and A. Alcala. 1994. "Sumilon Island Reserve: Twenty years of hopes and frustrations." *Naga, the ICLARM Quarterly* 17(3): 8–12.

[274] Ballantine, W.J. 1991. *Marine Reserves for New Zealand*. Leigh Laboratory Bulletin 25. University of Auckland.

[275] Shackell, Nancy L. and J.H. Willison. 1995. *Marine Protected Areas and Sustainable Fisheries*. Wolfville, NS: Science and Management of Protected Areas Association.

[276] Clark, A.T. 1996. "Refugia." Paper presented at the National Academy of Sciences International Conference on Ecosystem Management for Sustainable Marine Fisheries. February 19–24. Monterey, California.

[277] Ludwig, D.R., R. Hilborn, and C. Walters. 1993. "Uncertainty, resource exploitation, and conservation: Lessons from history." *Science* 260: 17 and 36.

[278] Parfit, M. and R. Kendrick. 1995. "Diminishing returns." *National Geographic* 188(5): 2–37.

[279] Safina, C. 1995. "The world's imperiled fish." *Scientific American* 273(5): 46–53.

[280] N.N. As originally published in the *New York Times* (Pauly, D. 2014. "Fishing more, catching less." *New York Times*, March 26). As was the case with the "Aquacalypse" article, the title is not one I have chosen; however, I kept it, as it pithily summarizes the fisheries issues of our time.

[281] N.N. See: Belhabib D., U.R. Sumaila, V.W.Y. Lam., D. Zeller, P. Le Billon, E.A. Kane, and D. Pauly. 2015. "Euro vs. Yuan: Comparing European and Chinese fishing access in West Africa." PLOS One 10(3): e0118351, and references therein.

[282] N.N. See: Pauly, D. and D. Zeller. 2016. "Catch reconstructions reveal that global marine fisheries catches are higher than reported and declining." *Nature Communications* 7. doi: 10.1038/ncomms10244.

[283] N.N. This contribution was originally published as Pauly, D. 1996. "Fleet-operational, economic, and cultural determinants of bycatch uses in Southeast Asia." In: *Solving Bycatch: Considerations for Today and Tomorrow*, 285–288. Sea Grant College Program, Report No. 96-03. University of Alaska Fairbanks and NOAA Fisheries. It acknowledged Sandra Gayosa (Philippines), Eny A. Buchary (Indonesia), Ratana Chuenpagdee (Thailand), and Alida Bundy (UK and Canada) for reading a draft and the improvements that ensued.

[284] Alverson, D.L., M. Freeberg, S.A. Murawski, and J.G. Pope. 1994. *A Global Assessment of Fisheries Bycatch and Discards*. FAO Fisheries Technical Paper 339. Rome: Food and Agriculture Organization of the United Nations.

[285] Dennett, D.C. 1995. *Darwin's Dangerous Idea: Evolution and the Meaning of Life*. New York: Simon & Schuster.

[286] Shindo, S. 1973. *A General Review of the Trawl Fishery and the Demersal Fish Stocks in the South China Sea*. FAO Fisheries Technical Paper 120. Rome: Food and Agriculture Organization of the United Nations.

[287] Pauly, D. and Chua Thia Eng. 1988. "The overfishing of marine resources: Socioeconomic background in Southeast Asia." AMBIO: A *Journal of Human Environment* 17(3): 200–206.

[288] Pauly, D. 1988. "Fisheries research and the demersal fisheries of Southeast Asia." In: *Fish Population Dynamics*, 2nd ed., edited by J.A. Gulland, 329–348. Chichester and New York: Wiley InterScience.

[289] Longhurst, A.R. and D. Pauly. 1987. *Ecology of Tropical Oceans*. San Diego: Academic Press.

[290] Sinoda, M., S.M. Tan, Y. Wanatabe, and Y. Meemeskul. 1979. "A method for estimating the best cod end mesh size in the South China Sea area." *Bulletin of the Choshi Marine Laboratory* 11: 65-80.

[291] Azhar, T. 1980. "Some preliminary notes on the bycatch of prawn trawlers off the west coast of Peninsular Malaysia." In: *Report of the Workshop on the Biology and Resources of Penaeid Shrimps in the South China Sea Area*, June 30-July 5, Part I, 64-69. Kota Kinabalu, Sabah, Malaysia.

[292] Sarjono, I. 1980. "Trawlers banned in Indonesia." ICLARM *Newsletter* 3(4): 3.

[293] Butcher, J.G. 1996. "The marine fisheries of the Western Archipelago: Toward an economic history." In: *Baseline Studies of Biodiversity: The Fish Resources of Western Indonesia*, edited by D. Pauly and P. Martosubroto. ICLARM Studies and Reviews 23.

[294] Mizutani, T., A. Kimizuka, K. Ruddle, and N. Ishige. 1987. "A chemical analysis of fermented fish products and discussion of fermented flavors in Asian cuisines." *Bulletin of the National Museum of Ethnology* 12(3): 801-864.

[295] Ruddle, K. 1986. "The supply of marine fish species for fermentation in Southeast Asia." *Bulletin of the National Museum of Ethnology* 11(4): 997-1036.

[296] N.N. The original reference for this statement was Alverson, D.L., M. Freeberg, S.A. Murawski, and J.G. Pope. 1994. *A Global Assessment of Fisheries Bycatch and Discards*. FAO Fisheries Technical Paper 339, Rome: Food and Agriculture Organization of the United Nations. This contribution, however, overestimated the extent of discarding; see: Pauly, D. and D. Zeller. 2016. "Catch reconstructions reveal that global marine fisheries catches are higher than reported and declining." *Nature Communications* 7. doi: 10.1038/ncomms10244.

[297] Pauly, D. 1988. "Fisheries research and the demersal fisheries of Southeast Asia." In: *Fish Population Dynamics*, 2nd ed., edited by J.A. Gulland, 329-348. Chichester and New York: Wiley InterScience.

[298] N.N. This contribution, reprinted here with permission, was originally published as Pauly, D. 1998. "Rationale for reconstructing catch time series." EC *Fisheries Cooperation Bulletin* 11(2): 4-7. It acknowledged the participants of the ACP-EU Course on Fisheries and Biodiversity Management, held in Port of Spain, Trinidad and Tobago, May 21-June 3, 1998, for their interest in the discussions that led to a draft and for their comments on it.

[299] N.N. Obviously, an update must here mention Google Earth, which can be used to count fishing boats, but also fishing installations such as weirs; see: Al-Abdulrazzak, D. and D. Pauly. 2013. "Managing fisheries from space: Google Earth improves estimates of distant fish catches." *ICES Journal of Marine Science*. doi:10.1093/icesjms/fst178.

[300] N.N. For an example of the kind of quite reliable information that can be elicited from old fishers see: Tesfamichael, D., T.J. Pitcher, and D. Pauly. 2014. "Assessing changes in fisheries using fishers' knowledge to generate long time series of catch rates: A case study from the Red Sea." *Ecology and Society* 19 (1): 18.

[301] N.N. A number of approximations of catch composition of the sort illustrated here can be averaged into a representative set of percentages, which can be applied to the catches of the relevant period. These percent catch compositions can be interpolated in time. E.g., for 1950–1954 with a composition of 40% groupers, 20% snappers, 10% grunts, and 30% other fish, and for these same groups in 1960–1964, a composition of 10%, 10%, 20%, and 60% respectively, the values for the intermediate period (1955–1959) can then be interpolated as 25% groupers, 15% snappers, 15% grunts, and 45% other fish.

[302] Kenny, J.S. 1955. "Statistics of the Port-of-Spain wholesale fish market." *Journal of the Agricultural Society* (June 1955): 267–272.

[303] King-Webster, W.A. and H.D. Rajkumar. 1958. *A Preliminary Survey of the Fisheries of the Island of Tobago*. Unpublished MS, 19 p. Port of Spain, Trinidad and Tobago: Caribbean Commission Central Secretariat.

[304] N.N. This contribution was originally published on June 8, 2016, with the title "A global, community-driven marine fisheries catch database," as part of a series produced by the *Huffington Post* in partnership with Ocean Unite, an initiative to unite and activate powerful voices for ocean-conservation action. The series was produced to coincide with World Ocean Day (June 8), as part of HuffPost's "What's Working" initiative, putting a spotlight on initiatives around the world that are solutions oriented.

[305] The data underlying all interactive graphic displays can be downloaded for further analysis.

[306] N.N. Therein, the definitions of large-scale ("industrial," often mislabeled "commercial") and small-scale (often mislabeled "traditional") are those prevailing in each maritime country. Governments tend to favor industrial fisheries, although it is the small-scale fisheries that meet most of the sustainability criteria.

[307] N.N. Pauly, D. and D. Zeller. 2016. "Catch reconstructions reveal that global marine fisheries catches are higher than reported and declining." *Nature Communications*, 7. doi: 10.1038/ncomms10244.

[308] N.N. This contribution is based on an essay solicited by Ms. Lucy Odling-Smee, a journalist working for *Nature*, who asked me "Why do we need

to know fisheries catches?" You never say no to a journalist working for *Nature*. However, I didn't know that my contribution (Pauly, D. 2013. "Does catch reflect abundance? Yes, it is a crucial signal." *Nature* 494: 303–306) would be part of a "debate" where someone would have the nerve to argue that we don't need to know fisheries catches. I wouldn't have bothered writing the essay then for the same reason that I don't debate creationists or climate denialists. Life is short, and there are much better ways to spend one's time and energy.

[309] Grainger, R.J.R. and S. Garcia. 1996. *Chronicles of Marine Fisheries Landings (1950–1994): Trend Analysis and Fisheries Potential*. FAO Fisheries Technical Paper 339. Rome: Food and Agriculture Organization of the United Nations.

[310] N.N. Froese, R. and K. Kesner-Reyes. 2002. *Impact of Fishing on the Abundance of Marine Species*. ICES CM 2002/L: 12. Copenhagen: International Council for the Exploration of the Sea.

[311] Worm, B., E.B. Barbier, N. Beaumont, J. Emmett Duffy, C. Folkes, B.S. Halpern, J.B.C. Jackson, H.K. Lotze, F. Micheli, S.R. Palumbi, E. Sala, A. Selkoe, J.J. Stachowicz, and R. Watson. 2006. "Impacts of biodiversity loss on ocean ecosystem services." *Science* 314: 787–790.

[312] N.N. See: Pauly, D. and D. Zeller. 2016. "Catch reconstructions reveal that global marine fisheries catches are higher than reported and declining." *Nature Communications*, 7. doi: 10.1038/ncomms10244.

[313] Kleisner, K., R. Froese, D. Zeller, and D. Pauly. 2013. "Using global catch data for inferences on the world's marine fisheries." *Fish and Fisheries* 14: 293–311.

[314] Froese, R., K. Kleisner, D. Zeller, and D. Pauly. 2012. "What catch data can tell us about the status of global fisheries." *Marine Biology* 159: 1283–1292.

[315] Walters, C.J. and J.J. Maguire. 1996. "Lessons for stock assessments from the northern cod collapse." *Reviews in Fish Biology and Fisheries* 6: 125–137.

[316] N.N. This work is now completed; it covered the Exclusive Economic Zone of 273 maritime countries (or part thereof) and their overseas territories, as well as the high seas; see: Pauly, D. and D. Zeller, eds. 2016. *Global Atlas of Marine Fisheries: A Critical Appraisal of Catches and Ecosystem Impacts*. Washington, DC: Island Press. While the fisheries catch data and derived indicators therein cover the years 1950 to 2010, updated versions of the data in this Atlas can be downloaded from http://www.seaaroundus.org.

[317] Zeller, D., S. Booth, G. Davis, and D. Pauly. 2007. "Re-estimation of small-scale fishery catches for U.S. flag-associated islands in the western Pacific: The last 50 years." *Fishery Bulletin* US 105: 266–277.

[318] Zeller, D., P. Rossing, S. Harper, L. Persson, S. Booth, and D. Pauly. 2011. "The Baltic Sea: Estimates of total fisheries removals 1950–2007." *Fisheries Research* 108: 356–363.

[319] N.N. This contribution, reprinted here with permission, was originally titled "Anecdotes and the shifting baseline syndrome of fisheries," and published as a one-page "Postscript" in *Trends in Ecology and Evolution* (1995, 10: 430). It was solicited by Robert May, who urgently needed some prose to complete the October 1995 issue of TREE. I quickly wrote a text based on some ideas that were buzzing in my head at the time, and Bob May was happy. Neither of us anticipated the success of this one-page article, which has garnered 1,676 citations on Google Scholar as of this writing (February 2018) and over 231,000 views of the corresponding TED Talk (see: http://www.ted.com/talks/daniel_pauly_the_ocean_s_shifting_baseline). More importantly, the paper may have inspired the founding of a new (sub)discipline—historical ecology—as documented, e.g., in the work of Sáenz-Arroyo, A., C. Robert, J. Torre, M. Cariño-Olvera, and R. Enríquez-Andrade. 2005. "Rapidly shifting environmental baselines among fishers of the Gulf of California." *Proceedings of the Royal Society of London B: Biological Sciences*, 272: 1957–1962, and especially in the following books: J.B.C. Jackson, K.E. Alexander, and E. Sala, eds. 2011. *Shifting Baselines: The Past and Future of Ocean Fisheries*. Washington, DC: Island Press; J.N. Kittinger, L. McClenachan, K.B. Gedan, and L.K. Blight, eds. 2014. *Marine Historical Ecology in Conservation: Applying the Past to Manage for the Future*. Berkeley: University of California Press; and D. Rost. 2014. *Wandel (v)erkennen: Shifting Baselines und die Wahrnehmung umweltrelevanter Veränderungen aus wissenssoziologischer Sicht*. Wiesbaden: Springer VS.

[320] The original version of this essay cited here an overestimate (30 million metric tons per year) based on the work of Alverson, D.L., M.H. Freeberg, J.G. Pope, and S.A. Murawski. 1994. *A Global Assessment of Fisheries Bycatch and Discards*. FAO Fisheries Technical Paper 339. Rome: Food and Agriculture Organization of the United Nations. This was deleted in view of the current estimate being lower (see: Zeller, D., T. Cashion, D.L.D. Palomares, and D. Pauly. 2018. "Global marine fisheries discards: A synthesis of reconstructed data." *Fish and Fisheries* 19(1): 30–39).

[321] The high estimate of 150 million metric tons was based on the approximate method that produced Figure 1 of Pauly, D., V. Christensen, S. Guénette, T.J. Pitcher, U.R. Sumaila, C.J. Walters, R. Watson, and D. Zeller. 2002. "Towards sustainability in world fisheries." *Nature* 418: 689–695. A much better estimate is 135 million metric tons (in 1996; see: http://www.seaaroundus.org), which is also well past previously

published estimates of global potential; see: Pauly, D. 1996. "One hundred million tonnes of fish, and fisheries research." *Fisheries Research* 25(1): 25-38.

322 Smith, T.D. 1994. *Scaling Fisheries*. Cambridge: Cambridge University Press.

323 Chapman, M.D. 1987. "Women fishing in Oceana." *Human Ecology* 15: 267-288.

324 N.N. See: Zeller, D., S. Harper, K. Zylich, and D. Pauly. 2015. "Synthesis of under-reported small-scale fisheries catch in Pacific-island waters." *Coral Reefs* 34: 25-39.

325 MacIntyre, F., K.W. Estep, and T.T. Noji. 1995. "Is it deforestation or desertification when we do it to the ocean?" *Naga, the ICLARM Quarterly* 18(3): 7-8.

326 Mowat, F. 1984. *Sea of Slaughter*. Atlantic Monthly Press, USA.

327 This contribution, reprinted here with permission, was originally published as Pauly, D. 2011. "On baselines that need shifting." *Solutions—for a Sustainable and Desirable Future* 2(1): 14. (Available from: http://www.thesolutionsjournal.com.)

328 N.N. The 48 contiguous states of the continental U.S.A.

329 N.N. The law in question is the Magnuson-Stevens Act, which, in the United States, mandates rebuilding of overexploited fish populations within 10 years to the level corresponding with "maximum sustainable yield." However, this requires assuming a population size (i.e., biomass level) that corresponds to the carrying capacity of the ecosystem (which is itself badly approximated by the size of an exploited population).

330 N.N. As the U.S. election of November 2016 demonstrated, fringe characters can be elected to the highest office, but whether they will remain there is an open question.

331 This contribution, reprinted here with permission, was originally published as Pauly, D. 2001. "Importance of the historical dimension in policy and management of natural resource systems." In: *Proceedings of the INCO-DEV International Workshop on Information Systems for Policy and Technical Support in Fisheries and Aquaculture*, edited by E. Feoli and C.E. Nauen, 5-10. ACP-EU Fisheries Research Report No. 8. It acknowledged Dr. Kim Bell, I.B.L. Smith Institute of Ichthyology, Grahamstown, South Africa, for his close reading of the draft and suggestions on how best to get its message across.

[332] In science, theories are not guesses, as they are in common usage of the word. On the contrary, theories, such as the Theory of Evolution by Natural Selection in biology, the Theory of Relativity in physics or the Plate Tectonic Theory in geology, are comprehensive articulations of the best, most reliable facts in a given discipline, together with their various interrelationships. New theories emerge slowly, but when they do, they allow a wide variety of disparate facts to be explained.

[333] Kuhn, T. 1962. *The Structure of Scientific Revolutions*. Chicago: University of Chicago Press.

[334] Gross, P.R. and N. Levitt. 1994. *Higher Superstition: The Academic Left and Its Quarrels with Science*. Baltimore: John Hopkins University Press.

[335] See contributions in: Pullin, R.S.V., R. Froese, and C.M.V. Casal, eds. 1999. *Proceedings of the Conference on Sustainable Use of Aquatic Biodiversity: Data, Tools and Cooperation*. Lisbon, Portugal, September 3-5, 1998. ACP-EU Fisheries Research Report No. 6.

[336] Pauly, D. 1996. "Biodiversity and the retrospective analysis of demersal trawl surveys: A programmatic approach." In *Baseline Studies in Biodiversity: The Fish Resources of Western Indonesia*, edited by D. Pauly and P. Martosubroto, 1-6. ICLARM Studies and Reviews 23.

[337] Froese, R. and D. Pauly, eds. 2000. *FishBase 2000: Concepts, Design and Data Sources*. Los Baños, Philippines: International Center for Living Aquatic Resources Management. (See also: http://www.fishbase.org.)

[338] N.N. Despite the enormous success of FishBase, which in early 2017 received 50 million "hits" per month (from half a million individual users), no specialist for groups other than fish followed our lead, even though we freely offered our advice, design, and software. Thus, with initial support from the Oak Foundation, we created SeaLifeBase (http://www.sealifebase.org), covering all multicellular marine organisms.

[339] N.N. Colléter, M., A. Valls, J. Guitton, D. Gascuel, D. Pauly, and V. Christensen. 2015. "Global overview of the applications of the Ecopath with Ecosim modeling approach using the EcoBase models repository." *Ecological Modelling* 24: 42-53. See: http://www.ecopath.org and http://sirs.agrocampus-ouest.fr/EcoBase/.

[340] N.N. Hoover, C., T.J. Pitcher, and V. Christensen. 2013. "Effects of hunting, fishing and climate change on the Hudson Bay marine ecosystem: I. Re-creating past changes 1970-2009." *Ecological Modelling* 24: 130-142.

[341] N.N. Sumaila, U.R., V.W.Y. Lam, F. Le Manach, W. Schwartz. and D. Pauly. 2016. "Global fisheries subsidies: An updated estimate." *Marine Policy* 69: 189-193. doi: 1016/j.marpol.2015.12.026.

[342] N.N. Since this was originally written, my positive impression of the potential of the Census of Marine Life was replaced by a sense that an excellent opportunity was, if not wasted, then strongly underutilized, as the Census ended up as a disparate assemblage of uncoordinated projects (fortunately costing far less than US$10 billion), leading in the aggregate to nothing resembling a "census"; see: Pauly, D. and R. Froese. 2010. "Account in the dark." *Nature Geoscience* 3(10): 662–663.

[343] N.N. This essay is reprinted here, with permission from the publisher, from Pauly, D. 2002. "Consilience in oceanographic and fishery research: A concept and some digressions." In: *The Gulf of Guinea Large Marine Ecosystem: Environmental Forcing and Sustainable Development of Marine Resources*, edited by J. McGlade, P. Cury, K.A. Koranteng, and N.J. Hardman-Mountford, 41–46. Amsterdam: Elsevier Science. It was based on a summary I gave of the workshop that led to the edited volume in which it was published. At the workshop held in Accra, Ghana, on July 27–29, 1998, in addition to presenting a case study on a Ghanaian coastal lagoon, I attempted to show the interrelationships between the multidisciplinary contributions presented at that workshop. I acknowledged Dr. Cornelia Nauen for discussions after this workshop, which was funded by the European Commission.

[344] Simonton, D.K. 1988. *Scientific Genius: A Psychology of Science*. Cambridge: Cambridge University Press.

[345] Pauly, D. 1994. "Resharpening Ockham's Razor." *Naga, the ICLARM Quarterly* 17(2): 7–8.

[346] Wilson, E.O. 1998. *Consilience: The Unity of Knowledge*. New York: Alfred A. Knopf.

[347] Alvarez, L.W., W. Alvarez, F. Asaro, and H.V. Michel. 1980. "Extraterrestrial cause for the Cretaceous-Tertiary extinction." *Science* 208: 1095–1106.

[348] Raup, D.M. 1986. *The Nemesis Affair: A Story of the Death of Dinosaurs and the Ways of Science*. New York: W.W. Norton.

[349] Gould, S.J. 1995. *Dinosaurs in a Haystack: Reflections in Natural History*. New York: Harmony Books. (The quote is on p. 152.)

[350] N.N. The resolution of this ancient conflict was essentially that evolution occurs when organisms are subjected to processes acting slowly and more or less uniformly over long periods, and that occasional catastrophes, such as a monster meteor slamming into Earth 65 million years ago, by annihilating huge fractions of ancient flora and fauna also had a major structuring role on the composition of the surviving flora and fauna, and hence on evolution's outcomes.

[351] N.N. Since this was written, I discovered that several countries have research groups and/or agencies to do just this.

[352] Cavalli-Sforza, L.L., A. Piazza, P. Menozzi, and J. Mountain. 1988. "Reconstruction of human evolution: Bringing together genetic, archaeological and linguistic data." *Proceedings of the National Academy of Science* 85(16): 6002–6006.

[353] Tort, P. 1996. "Sir William Thompson, Lord Kelvin, 1824–1907." In: *Dictionnaire du Darwinisme et de l'Evolution*, edited by P. Tort, 4281–4283. Paris: Presse Universitaire de France.

[354] Gilmont, R. 1959. *Thermodynamic Principles for Chemical Engineers*. New York: Prentice Hall. (The quote is on p. 146.)

[355] Sverdrup, H.U., M.W. Johnson, and R.H. Fleming. 1942. *The Oceans: Their Physics, Chemistry and General Biology*. New York: Prentice Hall.

[356] Bakun, A. 1996. *Patterns in the Ocean: Ocean Processes and Marine Population Dynamics*. California Sea Grant, La Jolla, California, and Centro de Investigaciones Biológicas del Noroeste, La Paz, Baja California Sur.

[357] Schrödinger, E. 1992. *What Is Life?* Cambridge: Cambridge University Press.

[358] Laevastu, T. and F. Favorite. 1977. *Preliminary Report on a Dynamic Numerical Marine Ecosystem Model (DYNUMES) for the Eastern Bering Sea*. U.S. National Marine Fisheries/Northwest and Alaska Fisheries Center, Seattle.

[359] Laevastu, T. and H.L. Larkins. 1981. *Marine Fisheries Ecosystems: Quantitative Evaluation and Management*. Farnham, Surrey: Fishing News Books.

[360] Polovina, J.J. 1984. "Model of a coral reef ecosystem I. The ECOPATH model and its application to French Frigate Shoals." *Coral Reefs* 3 (1): 1–11.

[361] Polovina, J.J. 1985. "An approach to estimating an ecosystem box model." *U.S. Fishery Bulletin* 83(3): 457–460.

[362] Christensen, V. and D. Pauly, eds. 1993. *Trophic Models of Aquatic Ecosystems*. ICLARM Conference Proceedings 26. Manila: International Center for Living Aquatic Resources Management.

[363] Pauly, D. 2002. "Spatial modelling of trophic interactions and fisheries impacts in coastal ecosystems: A case study of Sakumo Lagoon, Ghana." In: *The Gulf of Guinea Large Marine Ecosystem: Environmental Forcing and Sustainable Development of Marine Resources*, edited by J. McGlade, P. Cury, K.A. Koranteng, and N.J. Hardman-Mountford, xxxv and 289–296. Amsterdam: Elsevier Science.

[364] N.N. See: Okey, T. and D. Pauly. 1999. "A mass-balance trophic model of trophic flows in Prince William Sound: Decompartmentalizing ecosystem knowledge." In: *Ecosystem Approaches for Fisheries Management*, edited by S. Keller, 621–635. University of Alaska, Fairbanks.

[365] Hardman-Mountford, N.J. and J.M. McGlade. 2002. "Defining ecosystem structure from natural resource variability: Application of principal components analysis to remotely sensed sea surface temperatures." In: *The Gulf of Guinea Large Marine Ecosystem: Environmental Forcing and Sustainable Development of Marine Resources*, edited by J. McGlade, P. Cury, K.A. Koranteng, and N.J. Hardman-Mountford, 67–82. Amsterdam: Elsevier Science.

[366] Roy, C., P. Cury, P. Freon, and H. Demarq. 2002. "Environmental and resource variability off Northwest Africa." In: *The Gulf of Guinea Large Marine Ecosystem: Environmental Forcing and Sustainable Development of Marine Resources*, edited by J. McGlade, P. Cury, K.A. Koranteng, and N.J. Hardman-Mountford, 121–140. Amsterdam: Elsevier Science.

[367] Longhurst, A.R. 1998. *Ecological Geography of the Sea*. San Diego: Academic Press.

[368] Longhurst, A., R.S. Sathyendranath, T. Platt, and C.M. Caverhill. 1995. "An estimate of global primary production in the ocean from satellite radiometer data." *Journal of Plankton Research* 17: 1245–1271.

[369] Pauly, D. and V. Christensen. 1995. "Primary production required to sustain global fisheries." *Nature* 374: 255–257.

[370] Trites, A., V. Christensen, and D. Pauly. 1997. "Competition between fisheries and marine mammals for prey and primary production in the Pacific Ocean." *Journal of Northwest Atlantic Fishery Science* 22: 173–187.

[371] Pauly, D. 1995. "Anecdotes and the shifting baseline syndrome of fisheries." *Trends in Ecology and Evolution* 10(10): 430.

[372] Pauly, D. 1996. "Biodiversity and the retrospective analysis of demersal trawl surveys: A programmatic approach." In: *Baseline Studies in Biodiversity: The Fish Resources of Western Indonesia*, edited by D. Pauly and M. Martosubroto, 1–6. ICLARM Studies and Reviews 23. Manila: International Center for Living Aquatic Resources Management.

[373] Hilborn, R. and C.J. Walters. 1992. *Quantitative Fisheries Stock Assessment: Choice, Dynamics and Uncertainty*. New York: Chapman & Hall.

[374] Medawar, P.B. 1967. *The Art of the Soluble*. London: Methuen & Co.

[375] Gayet, M. 1996. "Humboldt, Friedrich Wilhelm Heinrich, Alexander von, 1769–1859." In: *Dictionnaire du Darwinisme et de l'Evolution*, edited by P. Tort, 2284–2287. Paris: Presse Universitaire de France.

[376] Conway, G.R. 1985. "Agroecosystem analysis." *Agricultural Administration* 20: 31–55.

[377] Note, incidentally that agroecosystems can also straightforwardly be described by trophic mass–balance models; see, e.g.: Dalsgaard, J.P.T. and R.T. Oficial. 1997. "A quantitative approach for assessing the productive performance and ecological contribution of smallholder farms." *Agricultural Systems* 55(4): 503–533.

[378] Pauly, D. and C. Lightfoot. 1992. "A new approach for analyzing and comparing coastal resource systems." *Naga, the ICLARM Quarterly* 15(3): 7–10.

[379] Nauen, C. "How can collaborative research be most useful to fisheries management in developing countries?" In: *The Gulf of Guinea Large Marine Ecosystem: Environmental Forcing and Sustainable Development of Marine Resources*, edited by J. McGlade, P. Cury, K.A. Koranteng, and N.J. Hardman-Mountford, 357–364. Amsterdam: Elsevier Science.

[380] N.N. Since this was originally written, coastal area management as described here has largely been replaced by conservation planning using formal software tools such as Marxan; see: Ball I.R, H.P. Possingham, and M. Watts. 2009. "Marxan and relatives: Software for spatial conservation prioritization." In: *Spatial Conservation Prioritization: Quantitative Methods and Computational Tools*, edited by A. Moilanen, K.A. Wilson, and H.P. Possingham, 185–195. Oxford: Oxford University Press, and other publications by Hugh P. Possingham and his associates.

[381] Eschmeyer, W.N., ed. 1998. *Catalog of Fishes*. Special publication, 3 volumes. San Francisco: California Academy of Sciences.

[382] Froese, R. and D. Pauly, eds. 1998. *FishBase 98: Concepts, Design and Data Sources*. Manila: International Center for Living Aquatic Resources Management. (See: http://www.fishbase.org for updates.)

[383] As might have been noted, several of my examples (Ecopath, SimCoast, FishBase) are products of projects (initially) supported by the European Commission, particularly by its (former) Directorate General concerned with development (i.e., VIII and XII INCO/DC), but not run by committees or university consortia. Perhaps this indicates that such projects, providing support to participants only if they buy into the strong concept underlying such ventures, are more effective than the usual collaborative projects, where the partners agree only to share the available funds.

[384] N.N. This contribution, reprinted here with permission, was originally published as Pauly, D. 2011. "Focusing one's microscope." *The Science Chronicles*

(The Nature Conservancy), January, 4-7. It was a quick response to an article suggesting that the phenomenon now widely known as "fishing down marine food webs," and documented by a multitude of authors in a multitude of cases throughout the world's oceans and in freshwaters in fact does not exist and is a figment of my imagination.

[385] Pauly, D. 2010. 5 Easy Pieces: The Impact of Fisheries on Marine Ecosystems. Washington, DC: Island Press.

[386] Branch, T.A., R. Watson, E.A. Fulton, S. Jennings, C.R. McGilliard, G.T. Pablico, D. Ricard, and S.R. Tracey. 2010. "The trophic fingerprint of marine fisheries." Nature 468: 431-435.

[387] N.N. See: Swartz, W., E. Sala, R. Watson, and D. Pauly. 2010. "The spatial expansion and ecological footprint of fisheries (1950 to present)." PLOS ONE 5(12): e15143. doi: 10.1371/journal.pone.0015143.

[388] N.N. See: Kleisner, K., H. Mansour, and D. Pauly. 2014. "Region-based MTI: Resolving geographic expansion in the Marine Trophic Index." Marine Ecology Progress Series 512: 185-199.

[389] Liang, C. and D. Pauly. 2017. "Fisheries impacts on China's coastal ecosystems: Unmasking a pervasive 'fishing down' effect." PLOS ONE 12(3): e0173296. doi:10.1371/journal.pone.0173296.

[390] N.N. This contribution (Pauly, D. 2014. "Homo sapiens: Cancer or parasite?" Ethics in Science and Environmental Politics 14(1): 7-10), perhaps born out of frustration with the continued devastation of the natural world, was originally part of a collection of essays in "Theme Section: The Ethics of Human Impacts and the Future of the Earth's Ecosystems," edited by D. Pauly and K. Stergiou. Ethics in Science and Environmental Politics 14(1). It is reprinted here with permission.

[391] Pauly, D., V. Christensen, S. Guénette, T.J. Pitcher, U.R. Sumaila, C.J. Walters, R. Watson, and D. Zeller. 2002. "Towards sustainability in world fisheries." Nature 418: 689-695.

[392] Pauly, D. 2009. "Aquacalypse now: The end of fish." The New Republic, October 7: 24-27.

[393] Pauly, D. 2009. "Beyond duplicity and ignorance in global fisheries." Scientia Marina 73: 215-223.

[394] Pauly, D. 2012. "Diagnosing and solving the global crisis of fisheries: Obstacles and rewards." Cybium 36: 499-504.

[395] Rapport, D.J. 2000. "Ecological footprints and ecosystem health: Complementary approaches to a sustainable future." Ecological Economics 32: 367-370.

396 Pavlikakis, G.E. and V.A. Tsihrintzis. 2003. "Integrating humans in ecosystem management using multi-criteria decision making." *Journal of the American Water Resource Association* 39: 277–288.

397 Hern, W.M. 1993. "Has the human species become a cancer on the planet? A theoretical view of population growth as a sign of pathology." *Current World Leaders* 36: 1089–1124.

398 MacDougall, A.K. 1996. "Human as cancer." *Wild Earth* 1996: 81–88.

399 Charles, A.T. 1995. "Fishery science: The study of fishery systems." *Aquatic Living Resources* 8: 233–239.

400 Berkes, F. 2004. "Rethinking community-based conservation." *Conservation Biology* 18: 621–630.

401 Jones, P.J.S. 2008. "Fishing industry and related perspectives on the issues raised by no-take marine protected area proposals." *Marine Policy* 32: 749–758.

402 Wells, S. 2004. *The Journey of Man: A Genetic Odyssey*. New York: Random House.

403 Stringer, C. 2011. *The Origin of Our Species*. London: Penguin Books.

404 Tattersall, I. 2009. "Human origins: Out of Africa." *Proceedings of the National Academy of Sciences USA* 106: 16018–16021.

405 Mellars, P. 2006. "Why did modern human populations disperse from Africa ca. 60,000 years ago? A new model." *Proceedings of the National Academy of Sciences USA* 103: 9381–9386.

406 Pinker, S. 2011. *The Better Angels of Our Nature: Why Violence Has Declined*. Penguin Books, London.

407 Wells, S. 2004. *The Journey of Man: A Genetic Odyssey*. Random House, New York.

408 Mellars, P. 2006. "Why did modern human populations disperse from Africa ca. 60,000 years ago? A new model." *Proceedings of the National Academy of Sciences USA* 103: 9381–9386.

409 Stringer, C. 2011. *The Origin of Our Species*. London: Penguin Books.

410 Oppenheimer, S. and M. Richards. 2001. "Fast trains, slow boats, and the ancestry of the Polynesian islanders." *Science Progress* 84: 157–181.

411 Alroy, J. 2001. "A multispecies overkill simulation of the end: Pleistocene megafaunal mass extinction." *Science* 292: 1893–1896.

412 Holdaway, R.N. and C. Jacomb. 2000. "Rapid extinction of the moas (Aves: Dinornithiformes): model, test, and implications." *Science* 287: 2250–2254.

413 Zimov, S.A. 2005. "Pleistocene park: Return of the mammoth's ecosystem." *Science* 308: 796–798.

[414] Flannery, T. 2002. *The Future Eaters: An Ecological History of the Australasian Lands and People.* New York: Grove Press.

[415] Montgomery, D.R. 2007. *Dirt: The Erosion of Civilizations.* Berkeley: University of California Press.

[416] Liebenberg, L. 2013. *The Origin of Science: The Evolutionary Roots of Scientific Reasoning and Its Implications for Citizen Science.* Cape Town: CyberTracker. (Available from: www.cybertracker.org/downloads/tracking/Liebenberg-2013-The-Origin-of-Science.pdf.)

[417] Purugganan, M.D. and D.Q. Fuller. 2009. "The nature of selection during plant domestication." *Nature* 457: 843–848.

[418] Davidson, D.J., J. Andrews, and D. Pauly. 2014. "The effort factor: Evaluating the increasing marginal impact of resource extraction over time." *Global Environmental Change* 25: 63–68.

[419] Ehrlich, P.R. 2014. "Human impact: The ethics of I=PAT." *Ethics in Science and Environmental Politics* 14: 11–18.

[420] Grimm, K.A. 2003. "Is earth a living system?" Geological Society of America *Abstracts with Programs* 35(6): 313. (Available from: https://gsa.confex.com/gsa/2003AM/finalprogram/abstract_62123.htm.)

[421] Tattersall, I. 2009. "Human origins: Out of Africa." *Proceedings of the National Academy of Sciences USA* 106: 16018–16021.

[422] Liebenberg, L. 2013. *The Origin of Science: The Evolutionary Roots of Scientific Reasoning and Its Implications for Citizen Science.* Cape Town: CyberTracker. (Available from: www.cybertracker.org/downloads/tracking/Liebenberg-2013-The-Origin-of-Science.pdf.)

[423] Murchison, E.P., C. Tovar, A. Hsu, H.S. Bender, P. Kheradpour, C.A. Rebbeck, D. Obendorf, C. Conlan, M. Bahlo, C.A. Blizzard, S. Pyecroft, A. Kreiss, M. Kellis, A. Stark, T.T. Harkins, J.A. Marshall Graves, G.M. Woods, G.J. Hannon, and A.T. Papenfuss. 2010. "The Tasmanian devil transcriptome reveals Schwann cell origins of a clonally transmissible cancer." *Science* 327: 84–87.

[424] Haldane, J.B.S. 1949. "Disease and evolution." *La Ricerca Scientifica* (Supplement) 19: 68–76.

[425] Hanski, I. 2014. "Biodiversity, microbes and human well-being." *Ethics in Science and Environmental Politics* 14: 19–25.

[426] Piketty, T. 2014. *Capital in the Twenty-first Century.* Cambridge, MA: Harvard University Press.

[427] Morowitz, H.J. 1992. *The Thermodynamics of Pizza: Essays on Science and Everyday Life.* New Brunswick, NJ: Rutgers University Press.

[428] Basu, K. 2014. "The whole economy is rife with Ponzi schemes." *Scientific American* 310: 70–75. (See also the essay titled "Aquacalypse Now: The End of Fish.")

[429] Sumaila, U.R. and C.J. Walters. 2005. "Intergenerational discounting: A new intuitive approach." *Ecological Economics* 52:135–142.

[430] Clark, C.W. 1973. "Profit maximization and the extinction of animal species." *Journal of Political Economy* 81: 950–961.

[431] Wikipedia, s.v. "Bernard Madoff." Accessed June 6, 2014, http://en.wikipedia.org/wiki/Bernard_Madoff.

[432] Ramankutty, N., L. Graumlich, F. Achard, D. Alves, A. Chhabra, B. DeFries, J. Foley, H. Geist, R. Houghton, K. Klein Goldewijk, E. Lambin, A. Millington, K. Rasmussen, R. Reid, and B.L. Turner. 2006. "Global land-cover change: Recent progress, remaining challenges." In *Land-Use and Land-Cover Change*, edited by E.F. Lambin and H. Geist, 9–39. Berlin: Springer.

[433] Pauly, D. 2009. "Aquacalypse now: The end of fish." *The New Republic*, October 7: 24–27.

[434] Toon, O.B., A. Robock, R.P. Turco, C. Bardeen, L. Oman, and G.L. Stenchikov. 2007. "Consequences of regional-scale nuclear conflicts." *Science* 315: 1224–1225.

[435] N.N. This contribution (based on Pauly, D. 2015. "Tenure, the Canadian tar sands and 'ethical oil.'" *Ethics in Science and Environmental Politics* 15(1): 55–57) was originally published in a collection of essays, "Academic freedom and tenure," edited by K.I. Stergiou and S. Somarakis. *Ethics in Science and Environmental Politics* 15(1). It is reprinted here with permission.

[436] Viens, A.M. and J. Savulescu. 2004. "Introduction to the Olivieri Symposium." *Journal of Medical Ethics* 30: 1–7.

[437] Hutchings, J.A., C.J. Walters, and R.L. Haedrich. 1997. "Is scientific inquiry incompatible with government control?" *Canadian Journal of Fisheries and Aquatic Science* 54: 1198–1210.

[438] Pannozzo, L. 2013. *The Devil and the Deep Blue Sea: An Investigation into the Scapegoating of Canada's Grey Seal*. Black Point, NS, and Winnipeg, MB: Fernwood Publishing.

[439] Walters, C.J. and J.J. Maguire. 1996. "Lessons for stock assessments from the northern cod collapse." *Reviews in Fish Biology and Fisheries* 6: 125–137.

[440] Hutchings, J.A., I.M. Côté, J.J. Dodson, I.A. Fleming, S. Jennings, N.J. Mantua, R.M. Peterman, B.E. Riddell, A.J. Weaver, and D.L.

VanderZwaag. 2012. *Sustaining Canadian Marine Biodiversity: Responding to the Challenges Posed by Climate Change, Fisheries, and Aquaculture.* RSC expert panel report prepared for the Royal Society of Canada, Ottawa.

[441] Pauly, D. 2007. "Obituary: Ransom Aldrich Myers (1954–2007)." *Nature* 447: 160.

[442] Government of Canada. 2006. *Communications Policy of the Government of Canada.* (Available from: https://www.tbs-sct.gc.ca/pol/doc-eng.aspx?id=12316.)

[443] O'Hara, K. 2010. "Canada must free scientists to talk to journalists." *Nature* 467: 501.

[444] Jones, N. 2013. "Canada to investigate muzzling of scientists." Newsblog, *Nature*, April 2. (Available from: http://blogs.nature.com/news/2013/04/canada-to-investigate-muzzling-of-scientists.html.)

[445] Turner, C. 2013. *The War on Science: Muzzled Scientists and Willful Blindness in Stephen Harper's Canada.* Vancouver, BC: Greystone Books.

[446] *The Globe and Mail.* 2013. "Editorial: Closing of research stations belies Ottawa's claim that it is protecting the environment." March 19.

[447] Bolen, M. 2014. "Tories accused of banning meteorologists from discussing climate change." *Huffington Post Canada*, May 30. (Available from: https://www.huffingtonpost.ca/2014/05/30/canadian-scientists-muzzled_n_5420607.html?utm_hp_ref=ca-harper-climate-change.)

[448] Bolen, M. 2014. "Mercer: Tories don't even understand the science they silence." *Huffington Post Canada*, November 20. (Available from: https://www.huffingtonpost.ca/2014/11/20/rick-mercer-science-conservatives-video_n_6192852.html.)

[449] Miller, K.M., S. Li, K.H. Kaukinen, N. Ginther, E. Hammill, J.M. Curtis, D.A. Patterson, T. Sierocinski, L. Donnison, P. Pavlidis, S.G. Hinch, K.A. Hruska, S.J. Cooke, K.K. English, and A.P. Farrell. 2011. "Genomic signatures predict migration and spawning failure in wild Canadian salmon." *Science* 331: 214–217.

[450] Morton, A., R. Routledge, and M. Krkosek. 2008. "Sea louse infestation in wild juvenile salmon and Pacific herring associated with fish farms off the east-central coast of Vancouver Island, British Columbia." *North American Journal of Fisheries Management* 28: 523–532.

[451] N.N. This is now shown to be the case: Di Cicco, E., H.W. Ferguson, A.D. Schulze, K.H. Kaukinen, S. Li, R. Vanderstichel, Ø. Wessel, E. Rimstad, I.A. Gardner, K.L. Hammel, and K.M. Miller. 2017. "Heart and skeletal muscle inflammation (HSMI) disease diagnosed on a British Columbia salmon farm through a longitudinal farm study." PLOS ONE 12(2):

e017471. doi: 10.1371/journal.pone.0171471. See also: Morton, A. 2017. "Mystery solved: This farm salmon disease is in BC." (Available from: http://alexandramorton.typepad.com/alexandra_morton/2017/02/mystery-solved-this-farm-salmon-disease-is-in-bc.html.)

[452] As evidenced in the film available from http://www.salmonconfidential.ca.

[453] Ecojustice. 2013. "Legal backgrounder: Fisheries Act." http://www.ecojustice.ca.

[454] Hutchings, J.A. and J.R. Post. 2013. "Gutting Canada's Fisheries Act: No fishery, no fish habitat protection." *Fisheries* 38: 497–501.

[455] Oreskes, N. and E.M. Conway. 2010. *Merchants of Doubt: How a Handful of Scientists Obscured the Truth on Issues from Tobacco Smoke to Global Warming.* New York: Bloomsbury Press.

[456] Oreskes, N. and E.M. Conway. 2014. *The Collapse of Western Civilization: A View from the Future.* New York: Columbia University Press.

[457] Stergiou, K.I. and A.C. Tsikliras, eds. 2013. "Global university rankings uncovered." *Ethics in Science and Environmental Politics* 13: 59–213.

[458] N.N. See: Boothe, P. 2015. "A word of advice for newly un-muzzled federal scientists." *Maclean's,* November 11. (Available from: www.macleans.ca/politics/ottawa/a-word-of-advice-for-canadas-newly-un-muzzled-federal-scientists/.)

[459] N.N. See: Owen, B. 2016. "Canada's government scientists get anti-muzzling clause in contract." *Science* 354: 358.

[460] N.N. This contribution was originally published as Pauly, D. 2008. "Worrying about whales instead of managing fisheries: A personal account of a meeting in Senegal." *Sea Around Us Project Newsletter,* May–June (47): 1–4.

[461] Kaczynski, V.M. and Fluharty, D.L. 2002. "European policies in West Africa: Who benefits from fisheries agreements?" *Marine Policy* 26: 75–93.

[462] Chavance, P., M. Ba, D. Gascuel, M. Vakily, and D. Pauly, eds. 2004. *Pêcheries Maritimes, Écosystèmes et Sociétés en Afrique de l'Ouest: Un Demi-siècle de Changement.* Actes du symposium international, Dakar, Sénégal, 24–28 juin, 2002. Office des Publications Officielles des Communautés Européennes, XXXVI, Collection des rapports de recherche halieutique ACP-UE 15.

[463] Alder, J. and U.R. Sumaila. 2004. "Western Africa: A fish basket of Europe past and present." *Journal of Environment and Development* 13: 156–178.

[464] N.N. Both had been PhD students of mine, and I was rather proud of their performance.

[465] N.N. This work, which built on earlier publications (Kaschner, K. and Pauly, D. 2005. "Competition between marine mammals and fisheries: Food for thought." In: *The State of Animals III: 2005*, edited by D.J. Salem and A.N. Rowan, 95–117. Washington, DC: Humane Society Press; and Swartz, W. and D. Pauly. 2008. "Who's eating all the fish? The food security rationale for culling cetaceans." Washington, DC: Humane Society of the United States), was subsequently summarized in Gerber, L., L. Morissette, K. Kaschner, and D. Pauly. 2009. "Should whales be culled to increase fishery yields?" *Science* 323: 880–881, with more detailed accounts in Morissette, L., V. Christensen, and D. Pauly. 2012. "Marine mammal impacts in exploited ecosystems: Would large-scale culling benefit fisheries?" PLOS ONE 7(9): e43966.

[466] N.N. Whale watching is also a vibrant industry along the coast of British Columbia, bringing more (and sustained) benefits that its earlier, particularly murderous, whaling industry ever did.

[467] I thank Dirk Zeller for reading and commenting on this new essay.

[468] Bourque B.J., B.J. Johnson, and R.S. Steneck. 2008. "Possible prehistoric fishing effects on coastal marine food webs in the Gulf of Maine." *Human Impacts on Ancient Marine Ecosystems: A Global Perspective*, edited by R. Torben and J.M. Erlanson, 165–185. Berkeley: University of California Press.

[469] Alexander K.E., W.B. Leavenworth, J. Cournane, A.B. Cooper, S. Claesson, S. Brennan, G. Smith, L. Rains, K. Magness, R. Dunn, and T.K. Law. 2009. "Gulf of Maine cod in 1861: Historical analysis of fishery logbooks, with ecosystem implications." *Fish and Fisheries* 10: 428–449.

[470] Pauly, D. 1986. "Problems of tropical inshore fisheries: Fishery research on tropical soft-bottom communities and the evolution of its conceptual base." In: *Ocean Yearbook 1986*, edited by E.M. Borgese and N. Ginsburg, 29–37. Chicago: University of Chicago Press.

[471] See: Finley, C. 2011. *All the Fish in the Sea: Maximum Sustainable Yield and the Failure of Fisheries Management*. Chicago: University of Chicago Press.

[472] It was published, instead in the house publication of the State Department; see: Chapman, W.M. 1949. *United States Policy on High Seas Fisheries*. Bulletin 20: 67–80. Washington, DC: Department of State.

[473] The relevant publications are (1) Schaefer, M.B. 1954. "Some aspects of the dynamics of populations important to the management of the commercial marine fisheries." *Bulletin of the Inter-American Tropical Tuna Commission* 1: 27–56; and (2) Schaefer, M.B. 1957. "A study of the

dynamics of populations of the fishery for yellowfin tuna in the eastern tropical Pacific Ocean." *Bulletin of the Inter-American Tropical Tuna Commission* 2: 227–268. It is a pity that the many scientists, historians of science, and fisheries pundits, always ready to criticize the MSY concept, hardly ever get that there was and still is a huge difference between the fakery that W.M. Chapman invented and the MSY and related concepts that M.B. Schaefer derived. Schaefer's MSY is based on the principle of density-dependent growth, one of the fundamental concepts of ecology, and also an essential component of evolution through natural selection, as proposed by Charles Darwin.

[474] Finley, C. 2017. *All the Boats in the Oceans: How Government Subsidies Led to Global Overfishing.* Chicago: University of Chicago Press.

[475] Weber, M.L. 2002. *From Abundance to Scarcity: A History of US Marine Fisheries Policy.* Washington, DC: Island Press.

[476] Interestingly, the members of these Regional Fishery Management Councils are exempted from conflict-of-interest regulations, which comes in handy since they are stacked with folks who have conflicts of interest; see: Okey, T.A. 2003. "Membership of the eight Regional Fishery Management Councils in the United States: Are special interests over-represented?" *Marine Policy* 27(3): 193–206.

[477] This section is adapted from Pauly, D. 2009. "Fish as food: A love affair, issues included." *Huffington Post*, November 12. (Available from: http://www.huffingtonpost.com/dr-daniel-pauly/fish-as-food-a-love-affai_b_354399.html.)

[478] Jenkins, D.J.A., J.L. Sievenpiper, D. Pauly, U.R. Sumaila, C.W.C. Kendall, and F.M. Mowat. 2009. "Are dietary recommendations for the use of fish oils sustainable?" *Canadian Medical Association Journal* 180(6): 633–637.

[479] See, e.g.: Burger, J. and M. Gochfeld. 2004. "Mercury in canned tuna: White versus light and temporal variation." *Environmental Research* 96: 239–249.

[480] See, e.g.: Hites R.A., J.A. Foran, D.O. Carpenter, M.C. Hamilton, B.A. Knuth, and S.J. Schwager. 2004. "Global assessment of organic contaminants in farmed salmon." *Science* 303: 226–229.

[481] See: (1) Jacquet, J. and D. Pauly. 2008. "Trade secrets: Renaming and mislabeling of seafood." *Marine Policy* 32: 309–318; (2) Cline, E. 2012. "Marketplace substitution of Atlantic salmon for Pacific salmon in Washington State detected by DNA barcoding." *Food Research International* 45: 388–393; and (3) Upton, H.F. 2015. "Seafood fraud." In: *Report for Congress*, April 7. Congressional Research Service.

[482] Pauly, D. and D. Zeller. 2016. "Catch reconstructions reveal that global marine fisheries catches are higher than reported and declining." *Nature Communications* 7. doi: 10.1038/ncomms10244.

[483] See: Jacquet, J., D. Pauly, D. Ainley, S. Holt, P. Dayton, and J. Jackson. 2010. "Seafood stewardship in crisis." *Nature* 467: 28-29.

[484] Because corporations are not people, contrary to various assertions to the contrary. Indeed, if they were people, they would be psychopaths, and their immediate institutionalization would be necessary to protect the public; see: Bakan, J. 2006. *The Corporation: The Pathological Pursuit of Profit and Power*. New York: Simon & Schuster.

[485] This is well documented in Jacquet, J. 2015. *Is Shame Necessary? New Uses for an Old Tool*. New York: Pantheon Books.

[486] See: Boonzaier, L. and D. Pauly. 2016. "Marine protection targets: An updated assessment of global progress." *Oryx, the International Journal of Conservation* 50: 27-35.

[487] Pala, C. 2017. "Four Pacific marine national monuments face threat under Trump order." *Earth Island Journal*, August 14. (Available from: http://www.earthisland.org/journal/index.php/elist/eListRead/four_marine_national_monuments_pacific_face_threat_trump/.)

[488] I thank Dirk Zeller for reading and commenting on this new essay.

[489] This account is adapted from Pauly, D. 2003. "Foreword/Avant-propos." In: *Une Taupe Chez les Morues: Halieuscopie d'un Conflit*, Bleus Marines Vol. I, by De Saint Pélissac, 5-9. Mississauga, ON: AnthropoMare.

[490] Indeed, one of my first papers was on the aquaculture of catfish in the southeastern United States (Pauly, D. 1974. "Report on the U.S. catfish industry: Development, research, production units, marketing and associated industries." In: *Neue Erkenntnisse auf dem Gebiet der Aquakultur*, edited by K. Tiews, 154-167. Arbeiten des Deutschen Fischereiverbandes, Heft 16. [In German]), and my master's thesis was about the aquaculture potential of blackchin tilapia, *Sarotherodon melanotheron*, in Ghana (Pauly, D. 1976. "The biology, fishery and potential for aquaculture of Tilapia melanotheron in a small West African lagoon." *Aquaculture* 7(1): 33-49). I also developed a method for the study of aquaculture operations (see, e.g.: van Dam, A.A. and D. Pauly. 1995. "Simulation of the effects of oxygen on food consumption and growth of Nile tilapia, *Oreochromis niloticus* (L.)." *Aquaculture Research* 26: 427-440), and I constantly monitored developments in aquaculture (see, e.g.: Stergiou, K.I., A.C. Tsikliras, and D. Pauly. 2009. "Farming up the Mediterranean food webs." *Conservation Biology* 23(1): 230-232).

491 Morton, A. 2002. *Listening to Whales: What the Orcas Have Taught Us*. New York: Ballantine Books.

492 I have even written about this; see: Pauly, D. 1987. "On using other people's data." *Naga, the ICLARM Quarterly* 11(1): 6-7. (Reprinted in: D. Pauly. 1994. *On the Sex of Fish and the Gender of Scientists: Essays in Fisheries Science*. Essay no. 19, 145-150. London: Chapman & Hall.)

493 One of her earliest documentations of this expertise is Morton, A., R. Routledge, C. Peet, and A. Ladwig. 2004. "Sea lice (*Lepeophtheirus salmonis*) infection rates on juvenile pink (*Oncorhynchus gorbuscha*) and chum (*Oncorhynchus keta*) salmon in the nearshore marine environment of British Columbia, Canada." *Canadian Journal of Fisheries and Aquatic Sciences* 61(2): 147-157.

494 See: Dill, L.M. and D. Pauly. 2004. "The [truth about the] science of fish lice." *Georgia Straight*, Dec. 16. Here, I put the words "truth about the" in square brackets because they were not part of the title of our submission to the *Georgia Straight*. Scientists don't argue about the "truth." Rather, they propose hypotheses, and then confront them with evidence, which then confirms or refutes the hypotheses.

495 Miller, K.M., S. Li, K.H. Kaukinen, N. Ginther, E. Hammill, J.M. Curtis, D.A. Patterson, K. Sierocinski, L. Donnison, P. Pavlidis, and S.G. Hinch. 2011. "Genomic signatures predict migration and spawning failure in wild Canadian salmon." *Science* 331: 214-217.

496 I deal with this episode, and generally with the muzzling of Canadian government scientists during the eight long years of the Harper government in: Pauly, D. 2015. "Tenure, the Canadian tar sands and 'ethical oil.'" *Ethics in Science and Environmental Politics* 15: 55-57 (see the essay titled "Academics in Public Policy Debates").

497 This was presumably due to a previously published paper showing that since 1900, the catches of the fisheries of British Columbia increasingly consist of smaller fishes, lower in the food web, i.e., that "fishing down" occurs in BC. See: Pauly, D., M.L.D. Palomares, R. Froese, P. Sa-a, M. Vakily, D. Preikshot, and S. Wallace 2001. "Fishing down Canadian aquatic food webs." *Canadian Journal of Fisheries and Aquatic Science* 58: 51-62.

498 Wallace, S. 1999. *Fisheries Impacts on Marine Ecosystems and Biological Diversity: The Role for Marine Protected Areas in British Columbia*. PhD thesis, Resource Management and Environmental Studies, University of British Columbia. Scott later documented the DFO's war of extermination against harmless basking sharks (see: Wallace, S. and Gisborne, B. 2006. *Basking Sharks: The Slaughter of BC's Gentle Giants*. Vancouver: Transmontanus) and now is a senior research scientist with the David Suzuki Foundation. In fact, lots of my ex-students work

in environmental NGOs. Note that this is not a "bug," and that their careers went astray: it is a feature.

[499] Dr. Cameron Ainsworth is now an associate professor at the University of South Florida.

[500] Ainsworth C. 2015. "British Columbia marine fisheries catch reconstruction, 1873 to 2011." *British Columbia Studies* 188: 81–89, 163.

[501] Incidentally, the same applies to Canada's arctic fisheries catches (mainly non-commercial), which are also ignored in federal data systems, and hence missing from Canada's FAO data; see: Zeller, D., S. Booth, E. Pakhomov, W. Swartz, and D. Pauly. 2011. "Arctic fisheries catches in Russia, USA and Canada: Baselines for neglected ecosystems." *Polar Biology* 34: 955–973.

[502] See: Government of Canada. "Open government." https://www.canada.ca/en/transparency/open.html.

[503] The results of these small consultancies were written up in Watkinson, S. and D. Pauly. 1999. *Changes in the Ecosystem of Rivers Inlet, British Columbia: 1950 vs. the Present*. A report to the David Suzuki Foundation, Vancouver, and in a master's thesis that studied the feasibility of setting up a marine protected area in Hecate Strait, BC, i.e., Beattie, A. 2002. *Optimal Size and Placement of Marine Protected Areas*. MSc thesis, Resource Management and Environmental Studies, University of British Columbia.

[504] Sarika was very productive; one of the most-cited papers emanating from her thesis evaluated the performance of the world's 18 Regional Fisheries Management Organizations (many of which have Canada as a member state); see: Cullis-Suzuki, S. and D. Pauly. 2010. "Failing the high seas: A global evaluation of regional fisheries management organizations." *Marine Policy* 34(5): 1036–1042.

[505] This was called the "Suzuki Diaries: Coastal Canada," directed by Caroline Underwood, and aired by the CBC, November 2009.

[506] Booth, S. and Paul Watts. 2007. "Canada's Arctic marine fish catch." In: *Reconstruction of Marine Fisheries Catches for Key Countries and Regions (1950–2005)*, edited by Zeller, D. and D. Pauly, 3–15. *Fisheries Centre Research Reports* 15(2). Vancouver, BC: University of British Columbia.

[507] Zeller, D., S. Booth, E. Pakhomov, W. Swartz, and D. Pauly. 2011. "Arctic fisheries catches in Russia, USA and Canada: Baselines for neglected ecosystems." *Polar Biology* 34(7): 955–973.

[508] See, e.g.: "Battle for the Arctic heats up." *CBC News*, February 27, 2009. (Available from: http://www.cbc.ca/news/canada/battle-for-the-arctic-heats-up-1.796010.)

[509] The RCMP claims that it killed the dogs for "health and safety" reasons, but they would say that, wouldn't they? (Here, I admit to paraphrasing the irresistible phrase spoken by the famous Mandy Rice-Davis.)

[510] Cheung, W.W.L., D. Pauly, and U.R. Sumaila. 2017. "Canadian fisheries and the world: The last 150 years." In: *Reflections of Canada: Illuminating our Possibilities and Challenges at 150 Years*, edited by P. Nemetz and M. Young, 237–243. Vancouver, BC: Peter Wall Institute for Advanced Students.

[511] See: Truth and Reconciliation Commission of Canada. 2015. *Honouring the Truth, Reconciling for the Future: Summary of the Final Report of the Truth and Reconciliation Commission of Canada*. (Available from: http://nctr.ca/reports.php.)

[512] Pauly, D. 2017. "Thoughts on UBC's Reconciliation Totem Pole." Sea Around Us Blog, April 13. (Available from: http://www.seaaroundus.org/thoughts-on-ubcs-reconciliation-totem-pole/.)

[513] Christensen, V. and D. Pauly. 1992. "The ECOPATH II: A software for balancing steady-state ecosystem models and calculating network characteristics." *Ecological Modelling* 61: 169–185.

[514] This is documented in: Pauly, D. and V. Christensen, eds. 1996. *Mass-Balance Models of North-Eastern Pacific Ecosystems. Fisheries Centre Research Reports* 4(1). Vancouver, BC: University of British Columbia.

[515] Walters, C., V. Christensen, and D. Pauly. 1997. "Structuring dynamic models of exploited ecosystems from trophic mass-balance assessments." *Reviews in Fish Biology and Fisheries* 7(2): 139–172.

[516] Pauly, D., V. Christensen, and C. Walters. 2000. "Ecopath, Ecosim and Ecospace as tools for evaluating ecosystem impact of fisheries." *ICES Journal of Marine Science* 57: 697–706.

[517] See: Watkinson, S. 2001. *Life after Death: The Importance of Salmon Carcasses in Watershed Function*. MSc thesis, Resource Management and Environmental Studies, University of British Columbia, Vancouver, BC.

[518] See the article by Nancy Baron. 2015. "Salmon Trees." *Hakai Magazine*, April 22. (Available from: https://www.hakaimagazine.com/features/salmon-trees/.) A more rigorous account can be found in: Reimchen, T.E., D.D. Mathewson, M.D. Hocking, J. Moran, and D. Harris. 2003. "Isotopic evidence for enrichment of salmon-derived nutrients in vegetation, soil, and insects in riparian zones in coastal British Columbia." In: *Nutrients in Salmonid Ecosystems: Sustaining Production and Biodiversity*, edited by J. Stockner, 59–70. American Fisheries Society Symposium 34. Bethesda, MD: American Fisheries Society.

[519] Jones, R., S. Watkinson, and D. Pauly. 2001. "Accessing traditional ecological knowledge of First Nations in British Columbia through local common names in FishBase." *Aboriginal Fisheries Journal/Newsletter of the BC Aboriginal Fish Commission* 7(1): Insert.

[520] See: http://www.fishbase.org. Incidentally, FishBase got a real boost from British Columbia, as the fish collection held by UBC's Zoology Department and now in the Beaty Biodiversity Museum was the first to be (electronically) incorporated in FishBase.

[521] Of the 50 graduate students whose MSc or PhD I supervised at UBC, two-thirds were or became Canadians.

[522] Polovina, J.J. 1984. "Model of a coral reef ecosystem." *Coral Reefs* 3(1): 1–11.

[523] Christensen, V. and D. Pauly. 1992. "The ECOPATH II: A software for balancing steady-state ecosystem models and calculating network characteristics." *Ecological Modelling* 61: 169–185.

[524] See, e.g.: Vega-Cendejas, M.E, F. Arreguin-Sanchez, and M. Hernández. 1993. "Trophic fluxes on the Campeche Bank, Mexico." In: *Trophic Models of Aquatic Ecosystems*, edited by V. Christensen and D. Pauly, 206–213. ICLARM Conference Proceedings 26. Manila: International Center for Living Aquatic Resources Management.

[525] But I retained my French citizenship...

[526] However, I had to demonstrate that when I went abroad, I did not always return to the same place. This required tracking ten years' worth of travels and hotel bills. Fortunately, I had kept those...

[527] N.N. I thank Dirk Zeller for reading and commenting on this essay, which owes its existence to two colleagues. The first, Jennifer Jacquet, argued that I shouldn't "waste" the opportunity of being invited as keynote speaker of the Marine Conservation Congress to be held in Washington, DC, in May 2009, by giving a conventional presentation with slides "that everybody knew." Rather, I should talk from the heart and tell why and how I related to ocean conservation. I followed that advice, and my presentation of May 20, given at the Smithsonian Institution in Washington, DC, was a huge success, if I may say so myself. The second colleague was Su Sponaugle, then the editor of the *Bulletin of Marine Science*, who suggested that I should write up the notes I assembled for my presentation and submit the resulting account to her journal. I did that, and the result was Pauly, D. 2011. "Toward a conservation ethic for the sea: Steps in a personal and intellectual odyssey." *Bulletin of Marine Science* 87: 165–175, which is reproduced here with permission. Its original acknowledgments included "J. Jacquet for insisting

that I should not, for my IMCC keynote, do a number-and-graph, PowerPoint-heavy lecture, but simply narrate my personal experience with, and my views on, marine conservation."

[528] N.N. See: Malakoff, D. 2002. "Going to the edge to protect the sea." *Science* 296: 458–461.

[529] See: Pauly, D. 1975. "On the ecology of a small West African lagoon." *Berichte der Deutschen Wissenschaftlichen Kommission für Meeresforschung* 24: 46–62; also: Pauly, D. 1976. "The biology, fishery and potential for aquaculture of *Tilapia melanotheron* in a small West African lagoon." *Aquaculture* 7(1): 33–49.

[530] van Banning, P. 1974. "A new species of *Paeonodes* (Therodamasidae, Cyclopoida), a parasitic copepod of the fish *Tilapia melanotheron* from the Sakumo-lagoon, Ghana, Africa." *Beaufortia* 22(286): 1–7.

[531] N.N. Wikipedia, s.v. "Proto-Human language." Accessed June 17, 2018, https://en.wikipedia.org/wiki/Proto-Human_language.

[532] N.N. Barkow, J.H., L. Cosmides, and J. Tooby, eds. 1995. *The Adapted Mind: Evolutionary Psychology and the Generation of Culture.* New York: Oxford University Press.

[533] N.N. See: Rees, W.E. 2000. "Patch disturbance, eco-footprints, and biological integrity: Revisiting the limits to growth (or why industrial society is inherently unsustainable)." *Ecological Integrity: Integrating Environment, Conservation, and Health* 1: 139–156.

[534] N.N. See: Erlandson, J.M., M.H. Graham, B.J. Bourque, D. Corbett, J.A. Estes, and R.S. Steneck. 2007. "The kelp highway hypothesis: Marine ecology, the coastal migration theory, and the peopling of the Americas." *The Journal of Island and Coastal Archaeology* 2(2): 161–174.

[535] N.N. Martin, P.S. 1984. "Prehistoric overkill: The global model." In: *Quaternary Extinctions: A Prehistoric Revolution*, edited by P.S. Martin and R.G. Klein, 354–403. Tucson: University of Arizona Press. See also: Alroy, J. 2001. "A multispecies overkill simulation of the end-Pleistocene megafaunal mass extinction." *Science* 292: 1893–1896.

[536] Montgomery, D.R. 2007. *Dirt: The Erosion of Civilizations.* Berkeley: University of California Press.

[537] N.N. This specific instance occurred at the 125th Annual Meeting of the American Fisheries Society, which took place August 26–September 1, 1995, in Tampa, Florida.

[538] N.N. See the essay titled "Focusing One's Microscope."

[539] N.N. Watson, R. and D. Pauly. 2001. "Systematic distortions in world fisheries catch trends." *Nature* 414: 534–536.

[540] N.N. Jackson, J.B.C., M.X. Kirby, W.H. Berger, K.A. Bjorndal, L.W. Botsford, B.J. Bourque, R. Cooke, J.A. Estes, T.P. Hughes, S. Kidwell, C.B. Lange, H.S. Lenihan, J.M. Pandolfi, C.H. Peterson, R.S. Steneck, M.J. Tegner, and R.R. Warner. 2001. "Historical overfishing and the recent collapse of coastal ecosystems." *Science* 293: 629–638.

[541] N.N. This applies especially to Myers, R.A. and B. Worm. 2003. "Rapid worldwide depletion of predatory fish communities." *Nature* 423: 280–283.

[542] N.N. See: Cashion, T., F. Le Manach, D. Zeller, and D. Pauly. 2017. "Most fish destined for fishmeal production are food-grade fish." *Fish and Fisheries* 18. doi: 10.1111/faf.12209.

[543] N.N. This is also available as a book: Brown, L.R. 2008. *Plan B 3.0: Mobilizing to Save Civilization* (substantially revised). New York: WW Norton & Company.

[544] This essay is based on a contribution solicited by the editor of the ICES *Journal of Marine Science*, Howard Bowman, as part of an autobiographical series by senior fisheries scientists. The original was published as Pauly, D. 2016. "Having to science the hell out of it." *ICES Journal of Marine Science* 73(9): 2156–2166. Reprinted by permission of Oxford University Press.

[545] Pauly, D. and I. Tsukayama, eds. 1987. *The Peruvian Anchoveta and Its Upwelling Ecosystem: Three Decades of Change*. ICLARM Studies and Reviews 15. Manila: International Center for Living Aquatic Resources Management.

[546] Swartz, W., E. Sala, R. Watson, and D. Pauly. 2010. "The spatial expansion and ecological footprint of fisheries (1950 to present)." *PLOS ONE* 5: e15143.

[547] Troadec, J.P., W.G. Clark, and J.A. Gulland. 1980. "A review of some pelagic fish stocks in other areas." *Rapports et Procès-Verbaux des Réunions/Conseil Permanent International pour l'Exploration de la Mer* 177: 252–277.

[548] Firth, R. 1946. *Malay Fishermen: Their Peasant Economy*. London: Keagan.

[549] Pauly, D. 2006. "Major trends in small-scale marine fisheries, with emphasis on developing countries, and some implications for the social sciences." *Maritime Studies* (MAST) 4: 7–22.

[550] Mannan, M. A. 1997. "Foreword." In: *Status and Management of Tropical Coastal Fisheries in Asia*, edited by G. Silvestre and D. Pauly. ICLARM Conference Proceedings 53. Manila: International Center for Living Aquatic Resources Management.

[551] Pauly, D. and P. Martosubroto. 1980. "The population dynamics of Nemipterus marginatus (Cuv. & Val.) off Western Kalimantan, South China Sea." *Journal of Fish Biology* 17: 263–273. Incidentally, the species in question was found later to have been Nemipterus thosaporni (see: http://www.fishbase.org), but this doesn't change the point made here.

[552] Weir, A. 2014. *The Martian*. New York: Crown Publishers.

[553] Pauly, D. 1978. "A discussion of the potential use in fish population dynamics of the interrelationships between mortality, growth parameters and mean environmental temperature in 122 fish stocks." Council Meeting 1978/G: 21, Demersal Fish Committee, International Council for the Exploration of the Sea.

[554] Pauly, D. 1980. "On the interrelationships between natural mortality, growth parameters and mean environmental temperature in 175 fish stocks." *Journal du Conseil International pour l'Exploration de la Mer* 39: 175–192.

[555] Pauly, D. 1984. *Fish Population Dynamics in Tropical Waters: A Manual for Use with Programmable Calculators*. ICLARM Studies and Reviews 8. Manila: International Center for Living Aquatic Resources Management.

[556] Pope, J. G. 1987. "Two methods for simultaneously estimating growth, mortality and cohort size parameters from time series of catch-at-length data from research vessel surveys." In: *Length-Based Methods in Fisheries Research*, edited by D. Pauly and G.R. Morgan, 103–111. ICLARM Conference Proceedings 13. Manila: International Center for Living Aquatic Resources Management.

[557] Sparre, P. 1987. "A method for the estimation of growth, mortality and gear selection/recruitment parameters from length-frequency samples weighted by catch per effort." In: *Length-Based Methods in Fisheries Research*, edited by D. Pauly and G.R. Morgan, 75–102. ICLARM Conference Proceedings 13. Manila: International Center for Living Aquatic Resources Management.

[558] Fournier, D.A., J.R. Sibert, J. Majkowski, and J. Hampton. 1990. "MULTIFAN: A likelihood-based method for estimating growth parameters and age composition from multiple length frequency data sets illustrated using data for southern bluefin tuna (Thunnus maccoyii)." *Canadian Journal of Fisheries and Aquatic Sciences* 47: 301–317.

[559] Fournier, D.A., J. Hampton, and J.R. Sibert. 1998. "MULTIFAN-CL: A length-based, age-structured model for fisheries stock assessment, with application to South Pacific albacore, Thunnus alalunga." *Canadian Journal of Fisheries and Aquatic Sciences* 55: 2015–2016.

[560] Munro, J.L. 2011. "Assessment of exploited stocks of tropical fishes." In: *Ecosystem Approaches to Fisheries: A Global Perspective*, edited by V. Christensen and J. Maclean, 145–170. Cambridge: Cambridge University Press.

[561] Pauly, D. and N. David. 1981. "ELEFAN I, a BASIC program for the objective extraction of growth parameters from length-frequency data." *Berichte der Deutschen Wissenschaftlichen Kommission für Meeresforschung* 28: 205–211.

[562] Gayanilo, F.C., P. Sparre, and D. Pauly. 1996. *The FAO-ICLARM Stock Assessment Tools (FiSAT) User's Guide.* FAO Computerized Information Series (Fisheries) No. 8. (Originally distributed with three diskettes.) Rome: Food and Agriculture Organization of the United Nations.

[563] Gayanilo, F.C., P. Sparre, and D. Pauly. 2005. *FAO-ICLARM Stock Assessment Tools II (FiSAT II). Revised Version User's Guide.* FAO Computerized Information Series (Fisheries) No. 8. (Distributed with one CD-ROM; http://www.fao.org/docrep/009/y5997e/y5997e00.htm. Rome: Food and Agriculture Organization of the United Nations. Also in French; Arabic (Yemen) translation, 2009, by A. Bakhraisa, with the support of EU Project EuropAid/126327/C/SER/YE.

[564] N.N. Pauly, D. and A. Greenberg, eds. 2013. *ELEFAN in R: A New Tool for Length-Frequency Analysis. Fisheries Centre Research Reports* 21(3). Vancouver, BC: University of British Columbia. The ELEFAN in R software can be downloaded from: https://github.com/AaronGreenberg/ELEFAN. The software requires a C++ compiler to build; Windows users may use Rtools.

[565] Pauly, D. 1981. "The relationships between gill surface area and growth performance in fish: A generalization of von Bertalanffy's theory of growth." *Berichte der Deutschen Wissenschaftlichen Kommission für Meeresforschung* 28: 251–282.

[566] Pauly, D. 2010. *Gasping Fish and Panting Squids: Oxygen, Temperature and the Growth of Water-Breathing Animals.* Excellence in Ecology. Book 22. Oldendorf/Luhe, Germany: International Ecology Institute.

[567] Cheung, W.W.L., J.L. Sarmiento, J. Dunne, T.L. Frölicher, V. Lam, M.L.D. Palomares, R. Watson, and D. Pauly. 2013. "Shrinking of fishes exacerbates impacts of global ocean changes on marine ecosystems." *Nature Climate Change* 3: 254–258.

[568] Pauly, D. 1997. "Geometrical constraints on body size." *Trends in Ecology and Evolution* 12: 442–443.

[569] Muir, B.S. and G.M. Hughes. 1969. "Gill dimensions for three species of tunny." *Journal of Experimental Biology* 51: 271–285.

[570] Hughes, G.M. 1970. "Morphological measurements on the gills of fishes in relation to their respiratory function." *Folia Morphologica (Praha)* 18: 78–95.

[571] Hughes, G.M. 1984. "Scaling of respiratory area in relation to oxygen consumption in vertebrates." *Experientia* 40: 519–524.

[572] De Jager, S. and W.J. Dekkers. 1975. "Relations between gill structure and activity in fish." *Netherlands Journal of Zoology* 25: 276–308.

[573] Palzenberger, M. and H. Pohla. 1992. "Gill surface area of water breather freshwater fishes." *Reviews in Fish Biology and Fisheries* 2: 187–192.

[574] Pauly, D. and W.W.L. Cheung. 2017. "Sound physiological knowledge and principles in modelling shrinking of fishes under climate change." *Global Change Biology* 24. doi: 10.1111/gcb.13831.

[575] All of these phenomena are explained in detail in Pauly, D. 2010. *Gasping Fish and Panting Squids: Oxygen, Temperature and the Growth of Water-Breathing Animals*. Excellence in Ecology. Book 22. Oldendorf/Luhe, Germany: International Ecology Institute.

[576] N.N. Calcareous concretions in the ear capsules of bony fishes used for perception of acceleration including gravity. Also called "ear bones" or "ear stones." These bones frequently show daily, seasonal, or annual checks, rings, or layers, which can be used to determine ages. Statoliths are similar to otoliths and fulfill the same functions in invertebrates, e.g., squids.

[577] See: Wikipedia, s.v. "Occam's razor." Accessed June 17, 2018, https://en.wikipedia.org/wiki/Occam%27s_razor.

[578] Cury, P. and D. Pauly. 2000. "Patterns and propensities in reproduction and growth of fishes." *Ecological Research* 15: 101–106.

[579] Bakun, A. 2011. "The oxygen constraint." In: *Ecosystem Approaches to Fisheries: A Global Perspective*, edited by V. Christensen and J. Maclean, 11–23. Cambridge: Cambridge University Press.

[580] Rhein, M., S.R. Rintoul, S. Aoki, E. Campos, D. Chambers, R.A. Feely, S. Gulev, G.C. Johnson, S.A. Josey, A. Kostianoy, C. Mauritzen, D. Roemmich, L.D. Talley, and F. Wang. 2013. "Observations: Ocean." In: *Climate Change 2013: The Physical Science Basis*, edited by T.F. Stocker, D. Qin, G.-K. Plattner, M. Tignor, S.K. Allen, J. Boschung, A. Nauels, Y.

Xia, V. Bex, and P.M. Midgley. Contribution of Working Group I to the Fifth Assessment Report of the Intergovernmental Panel on Climate Change. Cambridge and New York: Cambridge University Press.

[581] Keskin, C. and D. Pauly. 2014. "Changes in the 'Mean Temperature of the Catch': Application of a new concept to the North-eastern Aegean Sea." *Acta Adriatica* 55: 213–218.

[582] Tsikliras, A.C. and K.I. Stergiou. 2014. "Mean temperature of the catch increases quickly in the Mediterranean Sea." *Marine Ecology Progress Series* 515: 281–284.

[583] Cheung, W.W.L., R. Watson, and D. Pauly. 2013. "Signature of ocean warming in global fisheries catch." *Nature* 497: 365–368.

[584] Perry, A.L., P.J. Low, J.R. Ellis, and J.D. Reynolds. 2005. "Climate change and distribution shifts in marine fishes." *Science* 308: 1912–1915.

[585] Pauly, D. 2010. *Gasping Fish and Panting Squids: Oxygen, Temperature and the Growth of Water-Breathing Animals*. Excellence in Ecology. Book 22. Oldendorf/Luhe, Germany: International Ecology Institute.

[586] Cheung, W.W.L., J.L. Sarmiento, J. Dunne, T.L. Frolicher, V.W.Y. Lam, M.D. Palomares, and R. Watson. 2013. "Shrinking of fishes exacerbates impacts of global ocean changes on marine ecosystems." *Nature Climate Change* 3: 254–258.

[587] Cheung, W.W.L., V.W.Y. Lam, J.L. Sarmiento, K. Kearney, R. Watson, D. Zeller, and D. Pauly. 2010. "Large-scale redistribution of maximum fisheries catch potential in the global ocean under climate change." *Global Change Biology* 16: 24–35.

[588] Marr, J.C., ed. 1970. *The Kuroshio: A Symposium on the Japan Current*. Honolulu: East-West Center Press.

[589] Marr, J.C., D.K. Ohoeh, J. Pontecorvo, B.J. Rothschild, and A.R. Tuesing. 1971. *A Plan for Fishery Development in the Indian Ocean*. IOFC/DEV/71/1 Indian Ocean Fishery Commission. Rome: Food and Agriculture Organization of the United Nations and United Nations Development Programme.

[590] Pauly, D. 1979. *Theory and Management of Tropical Multispecies Stocks: A Review, with Emphasis on the Southeast Asian Demersal Fisheries*. ICLARM Studies and Reviews 1. Manila: International Center for Living Aquatic Resources Management.

[591] Cushing, D.H. 1982. "Review of 'Theory and Management of Tropical Multispecies Stocks.'" *Fisheries Research* 1: 182–184.

[592] Pullin, R.S.V. 2011. "Aquaculture up and down the food web." In: *Ecosystem Approaches to Fisheries: A Global Perspective*, edited by V. Christensen and J. Maclean, 89-119. Cambridge: Cambridge University Press.

[593] Christensen, V. and J. Maclean, eds. 2011. *Ecosystem Approaches to Fisheries: A Global Perspective*. Cambridge: Cambridge University Press.

[594] Pauly, D. 1976. "The biology, fishery and potential for aquaculture of Tilapia melanotheron in a small West African lagoon." *Aquaculture* 7: 33-49.

[595] Pauly, D. and A.N. Mines, eds. 1982. *Small-Scale Fisheries of San Miguel Bay, Philippines: Biology and Stock Assessment*. ICLARM Technical Report 7. Manila: International Center for Living Aquatic Resources Management.

[596] Smith, I.R. and A.N. Mines. 1982. *Small-Scale Fisheries of San Miguel Bay, Philippines: Economics of Production and Marketing*. ICLARM Technical Report 8. Manila: International Center for Living Aquatic Resources Management.

[597] Bailey, C. 1982. *Small-Scale Fisheries of San Miguel Bay, Philippines: Social Aspects of Production and Marketing*. ICLARM Technical Report 9. Manila: International Center for Living Aquatic Resources Management.

[598] Bailey, C. 1982. *Small-Scale Fisheries of San Miguel Bay, Philippines: Occupational and Geographic Mobility*. ICLARM Technical Report 10. Manila: International Center for Living Aquatic Resources Management.

[599] Smith, I.R., D. Pauly, and A.N. Mines. 1983. *Small-Scale Fisheries of San Miguel Bay, Philippines: Options for Management and Research*. ICLARM Technical Report 11. Manila: International Center for Living Aquatic Resources Management.

[600] Smith, I.R., and D. Pauly. 1983. *Small-Scale Fisheries of San Miguel Bay, Philippines: Resolving Multigear Competition in Nearshore Fisheries*. ICLARM Newsletter 6: 11-18. (Tagalog version ICLARM Translations 6, 1985; also available in the Bikol language.)

[601] Piketty, T. 2014. *Capital in the 21st Century*. Cambridge, MA: Harvard University Press.

[602] Cushing, D.H. 1988. "Review of the Peruvian anchoveta and its upwelling ecosystem: Three decades of change." *Journal du Conseil International Pour l'Exploration de la Mer* 44: 297-299.

[603] Pauly, D. and I. Tsukayama, eds. 1987. *The Peruvian anchoveta and Its Upwelling Ecosystem: Three Decades of Change*. ICLARM Studies and

Reviews 15. Manila: International Center for Living Aquatic Resources Management.

[604] Lindeman, R.L. 1942. "The trophic-dynamic aspect of ecology." Ecology 23: 399–418.

[605] Pauly, D. and V. Christensen. 2002. "Ecosystem models." In: *Handbook of Fish and Fisheries, Volume 2*, edited by P. Hart and J. Reynolds, 211–227. Oxford: Blackwell Publishing.

[606] Odum, E.P. 1969. "The strategy of ecosystem development." *Science* 104: 262–270.

[607] Pauly, D. 1975. "On the ecology of a small West African lagoon." *Berichte der Deutschen Wissenschaftlichen Kommission für Meeresforschung* 24: 46–62.

[608] Andersen, K.P. and E. Ursin. 1977. "A multispecies extension to the Beverton and Holt theory of fishing, with accounts of phosphorus circulation and primary production." *Meddelander fra Danmarks Fiskeri - og Havundersøgelser* 7: 319–435.

[609] Laevastu, T. and H.A. Larkins. 1981. *Marine Fisheries Ecosystem: Its Quantitative Evaluation and Management.* Farnham, Surrey, UK: Fishing News Books.

[610] Polovina, J.J. 1984. "Model of a coral reef ecosystem I. The ECOPATH model and its application to French Frigate Shoals." *Coral Reefs* 3(1): 1–11.

[611] Ulanowicz, R.E. 1986. *Growth and Development: Ecosystem Phenomenology.* New York: Springer Verlag.

[612] Pauly, D., M.L. Soriano-Bartz, and M.L.D. Palomares. 1993. "Improved construction, parametrization and interpretation of steady-state ecosystem models." In: *Trophic Models of Aquatic Ecosystems*, edited by V. Christensen and D. Pauly, 1–13. ICLARM Conference Proceedings 26. Manila: International Center for Living Aquatic Resources Management.

[613] Christensen, V. and D. Pauly. 1992. "The ECOPATH II: A software for balancing steady-state ecosystem models and calculating network characteristics." *Ecological Modelling* 61: 169–185.

[614] Christensen, V. and D. Pauly. 1992. *A Guide to the ECOPATH II Software System* (Version 2.1). ICLARM Software 6. Manila: International Center for Living Aquatic Resources Management. (Also available in French and Spanish.)

[615] Christensen, V. and D. Pauly, eds. 1993. *Trophic Models of Aquatic Ecosystems.* ICLARM Conference Proceedings 26. Manila: International Center for Living Aquatic Resources Management.

[616] Palomares, M.L.D., L. Morissette, A. Cisnero-Montemayor, D. Varkey, M. Coll, and C. Piroddi, eds. 2009. *Ecopath 25 Years Conference Proceedings: Extended Abstracts*. Fisheries Centre Research Reports 17(3). Vancouver, BC: University of British Columbia.

[617] Colléter, M., A. Valls, J. Guitton, D. Gascuel, D. Pauly, and V. Christensen. 2015. "Global overview of the applications of the Ecopath with Ecosim modeling approach using the EcoBase model repository." *Ecological Modelling* 302: 42–53.

[618] Palomares, M.L.D. and D. Pauly. 1998. "Predicting food consumption of fish populations as functions of mortality, food type, morphometrics, temperature and salinity." *Marine and Fisheries Research* 49: 447–453.

[619] Walters, C.J., V. Christensen, and D. Pauly. 1997. "Structuring dynamic models of exploited ecosystems from trophic mass-balance assessments." *Reviews in Fish Biology and Fisheries* 7: 139–172.

[620] Ibid.

[621] Walters, C.J., D. Pauly, and V. Christensen. 1998. "Ecospace: Prediction of mesoscale spatial patterns in trophic relationships of exploited ecosystems, with emphasis on the impacts of marine protected areas." *Ecosystems* 2: 539–554.

[622] Pauly, D., V. Christensen, and C. Walters. 2000. "Ecopath, Ecosim and Ecospace as tools for evaluating ecosystem impact of fisheries." *ICES Journal of Marine Science* 57: 697–706.

[623] Pauly, D. 2002. "Spatial modelling of trophic interactions and fisheries impacts in coastal ecosystems: A case study of Sakumo Lagoon, Ghana." In: *The Gulf of Guinea Large Marine Ecosystem: Environmental Forcing and Sustainable Development of Marine Resources*, edited by J. McGlade, P. Cury, K.A. Koranteng, and N.J. Hardman-Mountford, 289–296. Amsterdam: Elsevier Science.

[624] Colléter, M., A. Valls, J. Guitton, D. Gascuel, D. Pauly, and V. Christensen. 2015. "Global overview of the applications of the Ecopath with Ecosim modeling approach using the EcoBase model repository." *Ecological Modelling* 302: 42–53.

[625] See: National Oceanic and Atmospheric Administration (NOAA). "Top tens." (Available from: https://celebrating200years.noaa.gov/toptens.html.)

[626] Pauly, D. 1978. *A Preliminary Compilation of Fish Length Growth Parameters*. Berichte des Institut für Meereskunde an der Universität Kiel, No. 55.

[627] ICLARM. 1988. *ICLARM Five-Year Plan (1988–1992)*. Manila: International Center for Living Aquatic Resource Management.

[628] Froese, R. and D. Pauly, eds. 2000. *FishBase 2000: Concepts, Design and Data Sources.* Los Baños, Philippines: International Center for Living Aquatic Resource Management. (Available with four CD-ROMs; previous annual editions: 1996–1999.) Also available in Portuguese (1997, transl. of the 1996 edition), French (1998, updated transl. of the 1997 edition by N. Bailly and M.L.D. Palomares), and Chinese (2003, transl. of the 2000 edition by Kwang-Tsao Shao, Taiwan); updates in http://www.fishbase.org.

[629] McCall, R.A. and R. May. 1995. "More than a seafood platter." *Nature* 376: 735.

[630] Palomares, M.L.D. and D. Pauly, eds. 2015. *SeaLifeBase.* World Wide Web Electronic Publication. http://www.sealifebase.org.

[631] Palomares, M.L.D., D. Chaitanya, S. Harper, D. Zeller, and D. Pauly, eds. 2011. *The Marine Biodiversity and Fisheries Catches of the Pitcairn Group.* A report prepared for the Global Ocean Legacy Project of the Pew Environment Group by the Sea Around Us. Fisheries Centre, University of British Columbia, Vancouver, BC.

[632] Palomares, M.L.D., S. Harper, D. Zeller, and D. Pauly, eds. 2012. *The Marine Biodiversity and Fisheries Catches of the Kermadec Island Group.* A report prepared for the Global Ocean Legacy Project of the Pew Environment Group by the Sea Around Us. Fisheries Centre, University of British Columbia, Vancouver.

[633] N.N. See: Wikipedia, s.v. "Habilitation," https://en.wikipedia.org/wiki/Habilitation.

[634] Froese, R. and D. Pauly, eds. 2000. *FishBase 2000: Concepts, Design and Data Sources.* Los Baños, Philippines: International Center for Living Aquatic Resource Management. (Available with four CD-ROMs; previous annual editions: 1996–1999.) Also available in Portuguese (1997, transl. of the 1996 edition), French (1998, updated transl. of the 1997 edition by N. Bailly and M.L.D. Palomares), and Chinese (2003, transl. of the 2000 edition by Kwang-Tsao Shao, Taiwan); updates in http://www.fishbase.org.

[635] Pauly, D. and V. Christensen. 1995. "Primary production required to sustain global fisheries." *Nature* 374: 255–257. (Erratum in *Nature* 376: 279.)

[636] Pauly, D., V. Christensen, J. Dalsgaard, R. Froese, and F.C. Torres. 1998. "Fishing down marine food webs." *Science* 279: 860–863.

[637] Watson, R. and D. Pauly. 2001. "Systematic distortions in world fisheries catch trends." *Nature* 414: 534–536.

[638] Pauly, D., V. Christensen, S. Guénette, T.J. Pitcher, U.R. Sumaila, C.J. Walters, and R. Watson. 2002. "Towards sustainability in world fisheries." *Nature* 418: 689–695.

[639] Pauly, D., J. Alder, E. Bennett, V. Christensen, P. Tyedmers, and R. Watson. 2003. "The future for fisheries." *Science* 302: 1359-1361.

[640] Pauly, D. 2010. *5 Easy Pieces: How Fishing Impacts Marine Ecosystems*. Washington, DC: Island Press.

[641] Jackson, J.B.C., M.X. Kirby, W.H. Berger, K.A. Bjorndal, L.W. Botsford, B.J. Bourque, R. Cooke, J.A. Estes, T.P. Hughes, S. Kidwell, C.B. Lange, H.S. Lenihan, J.M. Pandolfi, C.H. Peterson, R.S. Steneck, M.J. Tegner, and R.R. Warner. 2001. "Historical overfishing and the recent collapse of coastal ecosystems." *Science* 293: 629-638.

[642] Myers, R.A. and B. Worm. 2003. "Rapid worldwide depletion of predatory fish communities." *Nature* 423: 280-283.

[643] Pauly, D. 1998. "Beyond our original horizons: The tropicalization of Beverton and Holt." *Reviews in Fish Biology and Fisheries* 8: 307-334.

[644] Pauly, D. 1998. "Why squid, though not fish, may be better understood by pretending they are." In: *Cephalopod Biodiversity, Ecology and Evolution*, edited by A.I.L. Payne, M.R. Lipinski, M.R. Clarke, and M.A.C. Roeleveld. *South African Journal of Marine Science* 20: 47-58.

[645] Pauly, D., A. Trites, E. Capuli, and V. Christensen. 1998. "Diet composition and trophic levels of marine mammals." *ICES Journal of Marine Science* 55: 467-481 (Erratum in *ICES Journal of Marine Science* 55: 1153, 1998); and Trites, A. and D. Pauly. 1998. "Estimating mean body mass of marine mammals from measurements of maximum body length." *Canadian Journal of Zoology* 76: 886-896.

[646] Pauly, D. 2004. *Darwin's Fishes: An Encyclopedia of Ichthyology, Ecology and Evolution*. Cambridge: Cambridge University Press.

[647] Many of these essays were reprinted in: Pauly, D. 1994. *On the Sex of Fish and the Gender of Scientists: Essays in Fisheries Science*. London: Chapman & Hall.

[648] Pauly, D. 1995. "Anecdotes and the shifting baseline syndrome of fisheries." *Trends in Ecology and Evolution* 10: 430.

[649] Engelhard, G.H., R.H. Thurstan, B.R. MacKenzie, H.K. Alleway, R. Colin, A. Bannister, M. Cardinale, M.W. Clarke, J.C. Currie, T. Fortibuoni, P. Holm, S.J. Holt, C. Mazzoldi, J.K. Pinnegar, S. Raicevich, F.A.M. Volckaert, E.S. Klein, and A.K. Lescrauwaet. 2015. "ICES meets marine historical ecology: Placing the history of fish and fisheries in current policy context." *ICES Journal of Marine Science* 73. doi:10.1093/icesjms/fsv219.

[650] Jackson, J.B.C., K.E. Alexander, and E. Sala, eds. 2011. *Shifting Baselines: The Past and the Future of Ocean Fisheries*. Washington, DC: Island Press.

[651] Kittinger, J.N., L. McClenachan, K.B. Gedan, and L.K. Blight, eds. 2014. *Marine Historical Ecology in Conservation: Applying the Past to Manage for the Future*. Berkeley: University of California Press.

[652] Rost, D. 2014. *Wandel (v)erkennen: Shifting Baselines und die Wahrnehmung umweltrelevanter Veränderung aus wissensoziologischer Sicht*. Wiesbaden: Springer vs.

[653] Bonfil, R., G. Munro, U.R. Sumaila, H. Valtysson, M. Wright, T. Pitcher, D. Preikshot, N. Haggan, and D. Pauly. 1998. "Impacts of distant water fleets: An ecological, economic and social assessment." In: *The Footprint of Distant Water Fleets on World Fisheries*, II–III. Endangered Seas Campaign, WWF International. Godalming, Surrey: World Wide Fund for Nature.

[654] Pauly, D. 1996. "A positive step: The Marine Stewardship Council initiative." *FishBytes, the Newsletter of the Fisheries Centre* 2: 1. Vancouver, BC: University of British Columbia.

[655] Jacquet, J., and D. Pauly. 2007. "The rise of seafood awareness campaigns in an era of collapsing fisheries." *Marine Policy* 31: 308–313.

[656] Jacquet, J., D. Pauly, D. Ainley, S. Holt, P. Dayton, and J. Jackson. 2010. "Seafood stewardship in crisis." *Nature* 467: 28–29.

[657] Pauly, D., V. Christensen, J. Dalsgaard, R. Froese, and F.C. Torres. 1998. "Fishing down marine food webs." *Science* 279: 860–863.

[658] Pauly, D. 2007. "The Sea Around Us Project: Documenting and communicating global fisheries impacts on marine ecosystems." *AMBIO: A Journal of Human Environment* 36: 290–295.

[659] Quinn, T.J. 2003. "Ruminations on the development and future of population dynamics models in fisheries." *Natural Resource Modeling* 16: 341–392.

[660] Sea Around Us. 2005. *Sea Around Us: A Five-Year Retrospective 1999 to 2004*. Sea Around Us Project, Fisheries Centre. Vancouver, BC: University of British Columbia.

[661] Pauly, D., V. Christensen, S. Guénette, T.J. Pitcher, U.R. Sumaila, C.J. Walters, and R. Watson. 2002. "Towards sustainability in world fisheries." *Nature* 418: 689–695.

[662] Pauly, D., J. Alder, E. Bennett, V. Christensen, P. Tyedmers, and R. Watson. 2003. "The future for fisheries." *Science* 302: 1359–1361.

[663] FAO. 2000. FISHSTAT Plus. *Universal Software for Fishery Statistical Time Series*. Version 2.3. Rome: Food and Agriculture Organization of the United Nations.

[664] Watson, R. and D. Pauly. 2001. "Systematic distortions in world fisheries catch trends." *Nature* 414: 534–536.

665 Pauly, D. and J. Maclean. 2003. *In a Perfect Ocean: The State of Fisheries and Ecosystems in the North Atlantic Ocean.* The State of the World's Oceans Series. Washington, DC: Island Press.

666 Christensen, V., S. Guénette, J.J. Heymans, C.J. Walters, R. Watson, D. Zeller, and D. Pauly. 2003. "Hundred-year decline of North Atlantic predatory fishes." *Fish and Fisheries* 4: 1–24.

667 Watson, R., A. Kitchingman, A. Gelchu, and D. Pauly. 2004. "Mapping global fisheries: Sharpening our focus." *Fish and Fisheries* 5: 168–177.

668 Belhabib, D., V. Koutob, A. Sall, V.W.Y. Lam, and D. Pauly. 2014. "Fisheries catch misreporting and its implications: The case of Senegal." *Fisheries Research* 151: 1–11.

669 Garibaldi, L. 2012. "The FAO global capture production database: A six-decade effort to catch the trend." *Marine Policy* 36: 760–768.

670 Zeller, D., S. Harper, K. Zylich, and D. Pauly. 2014. "Synthesis of underreported small-scale fisheries catch in Pacific-island waters." *Coral Reefs* 34: 25–39.

671 Zeller, D. and D. Pauly. 2005. "Good news, bad news: Global fisheries discards are declining, but so are total catches." *Fish and Fisheries* 6: 156–159.

672 Garibaldi, L. 2012. "The FAO global capture production database: A six-decade effort to catch the trend." *Marine Policy* 36: 760–768.

673 Pitcher, T.J., R. Watson, R. Forrest, H.P. Valtysson, and S. Guénette. 2002. "Estimating illegal and unreported catches from marine ecosystems: A basis for change." *Fish and Fisheries* 3: 310–339.

674 Zeller, D., R. Watson, and D. Pauly, eds. 2001. *Fisheries Impacts on North Atlantic Ecosystems: Catch, Effort and National/Regional Data Sets.* Fisheries Centre Research Reports 9(3). Vancouver, BC: University of British Columbia.

675 Zeller, D., S. Booth, E. Mohammed, and D. Pauly, eds. 2003. *From Mexico to Brazil: Central Atlantic Fisheries Catch Trends and Ecosystem Models.* Fisheries Centre Research Reports 11(6). Vancouver, BC: University of British Columbia.

676 Sea Around Us. 2010. *Sea Around Us: A Ten-Year Retrospective, 1999 to 2009.* Sea Around Us Project, Fisheries Centre. Vancouver, BC: University of British Columbia.

677 Pauly, D. and J. Maclean. 2003. *In a Perfect Ocean: The State of Fisheries and Ecosystems in the North Atlantic Ocean.* The State of the World's Oceans Series. Washington, DC: Island Press.

[678] Sea Around Us. 2005. *Sea Around Us: A Five-Year Retrospective 1999 to 2004.* Sea Around Us Project, Fisheries Centre. Vancouver, BC: University of British Columbia.

[679] Pauly, D. and D. Zeller. 2003. "The global fisheries crisis as a rationale for improving the FAO's database of fisheries statistics." In: *From Mexico to Brazil: Central Atlantic Fisheries Catch Trends and Ecosystem Models*, edited by D. Zeller, S. Booth, E. Mohammed, and D. Pauly, 1–9. *Fisheries Centre Research Reports* 11(6). Vancouver, BC: University of British Columbia.

[680] Zeller, D., S. Booth, and D. Pauly. 2005. *Reconstruction of Coral Reef and Bottom Fisheries Catches for U.S. Flag Islands in the Western Pacific, 1950–2002.* Report to the Western Pacific Regional Fishery Management Council, Honolulu.

[681] Zeller, D., S. Booth, G. Davis, and D. Pauly. 2007. "Re-estimation of small-scale fishery catches for U.S. flag-associated islands in the western Pacific: The last 50 years." *Fishery Bulletin* US 105: 266–277.

[682] Ibid.

[683] N.N. See, e.g.: Ruhlen, M. 1994. *On the Origin of Languages: Studies in Linguistic Taxonomy.* Stanford, CA: Stanford University Press.

[684] N.N. See also: Pauly, D. and D. Zeller, eds. 2016. *Global Atlas of Marine Fisheries: A Critical Appraisal of Catches and Ecosystem Impacts.* Washington, DC: Island Press.

[685] Pauly, D. and D. Zeller. 2016. "Catch reconstructions reveal that global marine fisheries catches are higher than reported and declining." *Nature Communications* 7. doi:10.1038/ncomms10244.

[686] Anticamara, J.A., R. Watson, A. Gelchu, and D. Pauly. 2011. "Global fishing effort (1950–2010): Trends, gaps, and implications." *Fisheries Research* 107: 131–136.

[687] Sumaila, U.R., V.W.Y. Lam, F. Le Manach, W. Schwartz, and D. Pauly. 2016. "Global fisheries subsidies: An updated estimate." *Marine Policy* 69. doi: 1016/j.marpol.2015.12.026.

[688] Pauly, D., W.W.L. Cheung, and U.R. Sumaila. 2015. "What are global studies?" Global Fisheries Cluster, Institute for the Oceans and Fisheries. Vancouver, BC: University of British Columbia. (Available from: http://www.global-fc.ubc.ca/what-are-global-studies/.)

[689] Froese, R., K. Kleisner, D. Zeller, and D. Pauly. 2012. "What catch data can tell us about the status of global fisheries." *Marine Biology* 159: 1283–1292.

[690] Kleisner, K., R. Froese, D. Zeller, and D. Pauly. 2013. "Using global catch data for inferences on the world's marine fisheries." *Fish and Fisheries* 14: 293–311.

[691] Pauly, D., V. Christensen, J. Dalsgaard, R. Froese, and F.C. Torres. 1998. "Fishing down marine food webs." *Science* 279: 860–863.

[692] Kleisner, K., H. Mansour, and D. Pauly. 2014. "Region-based MTI: Resolving geographic expansion in the Marine Trophic Index." *Marine Ecology Progress Series* 512: 185–199.

[693] Pauly, D. and A. Grüss. 2015. "Q&A: The present and the future of World and U.S. Fisheries—Interview with Daniel Pauly." *Fisheries* 40: 37–41.

[694] Pauly, D. and D. Zeller, eds. 2016. *Global Atlas of Marine Fisheries: A Critical Appraisal of Catches and Ecosystem Impacts*. Washington, DC: Island Press.

[695] Pauly, D., W.W.L. Cheung, and U.R. Sumaila. 2015. "What are global studies?" Global Fisheries Cluster, Institute for the Oceans and Fisheries. Vancouver, BC: University of British Columbia. (Available from: http://www.global-fc.ubc.ca/what-are-global-studies/.)

[696] Jack Marr and Ziad Shehadeh are now deceased.

[697] Acknowledging and adequately thanking the hundreds of people who positively influenced my career, including about seventy master's and PhD students who kept me on my toes, is impossible to do comprehensively. Here, I gratefully acknowledge A. Bakun, W. Cheung, V. Christensen, P. Cury, K.M. Freire, R. Froese, D. Gascuel, J. Jacquet, A.R. Longhurst, J. Maclean, C. Matthews, J. Mendo, J. Munro, C. Nauen, M. Palomares, M. Prein, R. Pullin, J. Reichert, I. Smith, K. Stergiou, U.R. Sumaila, M. Vakily, and D. Zeller. (J. Munro and I. Smith are now deceased.) The many more who should also have been listed here will no doubt mention it, and I hope to have the opportunity to make amends in the coming years.

[698] Diamond, J. 2005. *Collapse: How Societies Choose to Fail or Succeed*. New York: Viking.

[699] Given that in the United States, where most of Diamond's readers are, fish is consumed mainly in restaurants or as take-away food (e.g., fish burgers), the impact of the MSC on the choice of fish consumed in that country will remain minimal.

[700] Sumaila, U.R., W.W.L. Cheung, A. Dyck, K.M. Gueye, L. Huang, V. Lam, D. Pauly, U.T. Srinivasan, W. Swartz, R. Watson, and D. Zeller. 2012. "Benefits of rebuilding global marine fisheries outweigh costs." PLOS ONE 7(7): e40542.

[701] Stern, N.H. 2007. *The Economics of Climate Change: The Stern Review.* Cambridge and New York: Cambridge University Press.

[702] Adapted from the "Acronym and Glossary" section in: *Global Atlas of Marine Fisheries: A Critical Appraisal of Catches and Ecosystem Impacts,* edited by D. Pauly and D. Zeller, 459–477. Washington, DC: Island Press. Copyright © 2016. Used by permission of Island Press, Washington, DC.

INDEX

academic tenure, 127, 130
Academy of Science, 130
access agreements, 7–8, 71, 227n144
acidification, ocean, 27–28, 226n139
Africa: distant-water fleets in, 7, 35, 36, 133; fisheries development projects, 36; jellyfish outbursts, 27; seafood imports from, 71; SIAP project on catch data, 133–34; support for whales, 135–36; whales blamed for fisheries decline, 8, 132, 133, 134–35, 136
agriculture, 123, 162
Ainsworth, Cameron, 148, 261n499
Alaska, salmon fisheries, 138–39
Alexander, K.E., 186
Allsopp, W.H.L., 3
Alvarez, Luis, 109
Alvarez, Walter, 109
Amason, Ragnar, 50, 53, 54, 235n222
American Fisheries Society, Canadian Chapter, 155
American Samoa, 38–39
anchoveta fishery (Peru), 3, 169, 179–80
anecdotes. See historical observations and data
Annala, John, 51, 52
anthropology, 34, 45–46, 161
aquaculture: aeration of ponds and fish growth, 175; author's studies on, 178, 260n490; in China, 10, 12, 29; salmon, 129, 142, 146–48; tuna, 26–27, 143; unsustainable nature of, 10, 29–30, 143
Arctic fisheries, 149–51, 261n501
Argentina, 136

Arntz, Wolf, 179
Asian Development Bank, 74, 159, 170
Asimov, Isaac, 118
astronomy, 96

Bakun, Andy, 176
Baranov, F.I., 56, 64
baselines, shifting: author's contribution to literature, 186–87; explanation of, 95, 99–100, 226n131; in fisheries, 24, 100; positive shifts, 100–102; prevention using consilience, 112–13; prevention using historical and recovered knowledge, 96–98, 106–7
bathymetric expansion, 5–6
Beverton, Ray, 58, 172
Bijagós Archipelago (Guinea-Bissau), 44
biomass, fish: ocean acidification and, 28; reductions of, 6, 9, 47, 97
bioquads (occurrence records), 104–5
Blight, L.K., 186
Bolinao reef fishery (Philippines), 44
Booth, Shawn, 150
Brander, Keith, 173
Britain, 5, 9
British Columbia: Atlantic salmon aquaculture, 129, 146–48; catch reconstruction, 148; "fishing down" in, 261n497; whale watching, 257n466
Brown, Lester, 166
Burma (Myanmar), 74
Bush, George W., 144
bycatch, 3, 37, 69, 73, 75–76, 94

Index

California: sardine fishery, 2
Canada: Arctic fisheries, 149–51, 261n501; author's engagement with history, 152; commitment to transparency, 131, 149; fisheries crisis in, 60; fisheries management in, 141; Inuit policies, 151, 262n509; oil sands, 129–30; resistance to fisheries propaganda, 145–46; suppression of scientific findings, 127–28, 128–29, 129–30; turbot war, 5, 216n23. *See also* British Columbia; cod fishery, northern; Department of Fisheries and Oceans
Canadian Association of University Teachers, 130
Canadian Society for Ecology and Evolution, 131
Canadian Society of Zoologists, 131
cancer: economy as, 125–26; *Homo sapiens* as, 124–25
Caribbean, 8, 36, 44, 80–81, 85
Carson, Rachel, 157, 164, 188
catch data: actual catch estimates, 23, 94, 225n129, 245n321; Chinese over-reporting, 12, 93, 165, 190; debate over usefulness of, 89–90, 92; importance of, 86, 91–92, 93; national catch statistics, 214n3; need for better collection, 93; problems with, 2, 37, 80, 86, 191; Sea Around Us analysis project, 93, 164–65, 190; Sea Around Us database, 38, 87–88, 193–94, 233n216; short-term data collection, 80; SIAP project for Northwest Africa, 133–34; small-scale fisheries underestimation, 37–39, 71–72; *State of Fisheries and Aquaculture* (FOA), 214n5; stock assessments, 51–52, 89, 92, 179; stock-status plots, 90–91, 194; under-reporting, 13–14, 93; *Yearbook of Fisheries and Aquaculture Statistics* (FOA), 1–2, 89
catch reconstruction: Arctic fisheries, 149–51; in British Columbia, 148; data sources, 82, 104–6; historical observations for preventing baseline shifts, 96–98, 106–7; importance of, 80–81; process of, 83–84, 242n301; psychological aspect, 81–82; Sea Around Us project, 14, 191–93; Trinidad and Tobago example, 85
catch shares, 48. *See also* individual transferable quotas
cells, 116–17
Census of Marine Life, 106, 247n342
change, types of, 15–16
Chapman, Wilbert M., 138, 258n473
China: approach to fisheries, 223n108; aquaculture, 10, 12, 29; catch over-reporting, 12, 93, 165, 190; distant-water fleets, 7–8, 71; masking of fisheries decline, 119; seafood market, 165
Christensen, Villy, 96, 153, 164, 181
Christy, Francis, 49, 54
climate research, 104. *See also* global warming
coastal area management (CAM), 112, 113–14, 250n380
cod fishery, northern: author's experience with, 158, 187; collapse of, 4, 7, 23, 52, 92, 128; history of, 137–38; traditional refuge of, 65
cod war, 5
Cohen Commission, 129, 148
Collapse (Diamond), 196
co-management, 63–64
Common Fisheries Policy (CFP), 68–70
community-based approaches, 63–64
conservation, 20, 60, 69. *See also* consumer awareness campaigns; marine reserves
consilience: coastal area management example, 113–14; Cretaceous-Tertiary (K-T) extinction example, 108–9; ecosystem modeling example, 111–12; explanation of, 108, 109–10, 115; FishBase example, 114; *Homo sapiens* example, 108, 109–10; language and, 113; mass-balance

example, 110–11; vs. multidisciplinary work, 112; shifting baseline prevention example, 112–13
Consultative Group on International Agricultural Research (CGIAR), 184–85
consumer awareness campaigns, 11, 30–31, 143–44, 166. *See also* Marine Stewardship Council
consumption, of fish, 12, 21–22, 142–43
Convention on Biological Diversity, 19
coral-reef fisheries, 64, 97
corporations, 144, 259n484
Cretaceous-Tertiary (K-T) extinction, 108–9
Cullis-Suzuki, Sarika, 149, 261n504
Cushing, David, 172, 177, 180

Darwin, Charles, 186, 188
David, Noel, 174
David Suzuki Foundation, 149
Dawkins, Richard, 48
demersal realm, 5, 6, 57. *See also* trawl industry
Department of Fisheries and Oceans (DFO), 128, 129, 146, 147–48, 149, 261n498
Diamond, Jared: *Collapse*, 196
distant-water fleets, 7–8, 10, 35, 36, 38, 45, 71, 133
The Distinguished Gentleman (film), 137

ecology-based approaches, 64–66
economy, 26, 125–26
Ecopath: for data recovery, 105; development and dissemination, 154–55, 181–82, 186, 250n383; mass-balance approach, 111; reactions to, 174; Strait of Georgia project, 152–53
Ecopath with Ecosim (EwE), 153, 182
ecosystem modeling, 111–12, 153, 180–81. *See also* Ecopath
egg production, 56
Egypt, 162–63
ELEFAN (Electronic Length Frequency Analysis), 159, 173–74

environment: blamed for fishery collapse, 3; human approaches to, 166–68; public as stakeholders, 189–90
ethical issues, 59–60, 142–43
European Union: Common Fisheries Policy for fisheries renewal, 68–70; distant-water fleets, 7, 71, 133; fish consumption and seafood imports, 12, 70, 165; fisheries management in, 141; individual transferable quotas and, 53
evolution, 109, 248n350
Exclusive Economic Zones (EEZs), 7–8, 18, 31–32, 139, 227n144
expansion: overview, 21–22, 65, 165; bathymetric, 5–6; geographic, 4–5; as masking factor, 119; taxonomic, 6
extinction, fisheries, 46

fish-aggregating devices (FADs), 5
FishBase: consilience and, 114; development of, 105, 182–84, 250n383, 263n520; Indigenous fish names project and, 154; reactions to, 174; Sea Around Us database and, 87–88; success of, 184, 246n338
fisheries: current trends, 9–11, 94, 165–66; expansion trends, 4–6, 21–22, 65, 119, 165; future possibilities, 197; growth post-WWII, 2, 35–36; as Ponzi scheme, 21–23; public as stakeholders, 189–90; renewal of, 14–15. *See also* catch data; catch reconstruction; fisheries collapse; fisheries management; fisheries science
Fisheries Act, 130
fisheries biologists, 16–17, 25, 34, 99. *See also* fisheries science
Fisheries Centre (UBC), 153, 154–55, 185, 186
fisheries collapse: overview, 2–3, 23–24, 193; economic impacts, 26; masking factors, 12–13, 119; oceans' responses to, 26–28; realization of global depletion,

24–25; total collapse prediction, 24, 91, 225n130; toxic triad for, 3–4; types of issues from, 59–60 fisheries economists, 28–29, 45. See also fisheries science
fisheries extinction, 46
fisheries management: co-management, 63–64; Common Fisheries Policy (EU), 68–70; community-based approaches, 63–64; ecology-based approaches, 64–66; future possibilities, 18–19; government regulation, 28–29, 31–32; impacts of low fishing pressure, 56–57; Magnuson-Stevens Act (U.S.), 68, 140, 141, 245n329; market-based approaches, 62–63; modern governance approach, 64; need for engagement with local communities, 55, 60–61, 66–67; privatization trend, 55–56; responsibility for, 34; rule-based vs. ministerial discretion, 141; social scientists and, 34, 36–37, 38, 39, 45–46. See also marine reserves; small-scale fisheries; individual transferable quotas
fisheries science: change needed in, 16–18; future possibilities, 18–19; loss of credibility, 59; 1970s transition, 169–70; responses to fisheries crisis, 25–26, 95; shifting baseline syndrome, 95, 99–100; traditional approach to fisheries, 16, 189. See also fisheries biologists; fisheries economists
fish farming. See aquaculture
fish growth, 174–77, 226n138
fishing down marine food webs, 117–18
fishing-industrial complex, 22–23, 26
fishing rights, 18, 28–29, 45. See also individual transferable quotas
fishing vessels, 9
fish sauces, 76–78
Food and Agriculture Organization (FAO): author's attempt to join, 173; ELEFAN and, 174; *State of Fisheries and Aquaculture*, 214n5; voluntary guidelines for small-scale fisheries, 233n216; *Yearbook of Fisheries and Aquaculture Statistics*, 1–2, 89. See also catch data
Fraser, John, 145–46
Froese, Rainer, 90, 183
funding, for research, 103–4

Gedan, K.B., 186
geographic expansion, 4–5
German international development agency (GTZ), 158, 170
Germany, 10–11
Ghana: fisheries development projects, 36; Sakumo Lagoon fishery, 158, 178, 180–81, 182
Global Atlas of Marine Fisheries (Sea Around Us), 194, 243n316
global warming, 13, 15, 27–28, 96, 130, 197, 226n138
Google Earth, 242n299
Grafton, R. Quentin, 49, 50
Great Britain, 5, 9
Guinea-Bissau: Bijagós Archipelago, 44
Gulf of Mexico, 27, 138
Gulland, John, 3, 170, 173, 181

Hansen, Victor, 157
health issues, from fish, 142
Hempel, Gotthilf, 151, 170, 172, 195
herring fishery, 3
Hilborn, Ray, 13
historical ecology, 186
historical linguistics, 192
historical observations and data: consilience and, 112–13; for preventing baseline shifts, 96–98, 106–7; role in research, 102–3; strategies for recovering data, 82, 104–6. See also catch reconstruction
Holt, S.J., 58
Homo sapiens: introduction, 121; agriculture and economic development, 123–24, 125–26, 162; approaches to environment,

166–68; as "cancer of the earth," 124–25; as consilience example, 108, 109–10; environmental impacts, 161–63; evolution and global expansion, 122–23, 161–62; as parasites, 126; as "part of the ecosystem," 121–24
Hooke, Robert, 116
Hudson River sturgeon, 24
humans. See *Homo sapiens*
Humboldt, Alexander von, 114

Iceland, 5, 50, 53–54, 137, 235n222
individual transferable quotas (ITQs): introduction, 45, 48–49; concentration concern, 50; corporate ownership concern, 49; development of, 49, 95; electorate control over, 50; in Iceland, 50, 53–54, 235n222; impacts of, 62–63; meme metaphor for spread of, 48, 53–54; monitoring needs, 52; in New Zealand, 235n225; perceptions of, 28–29, 62; science and research needs, 51–52; for small-scale fisheries, 50–51; tropical and subtropical multispecies fisheries and, 54
Indonesia, 41, 42, 73, 74, 76, 158–59, 170–71
inshore fisheries, 38
Institute for the Oceans and Fisheries (UBC), 186. *See also* Fisheries Centre
International Center for Living Aquatic Resource Management (ICLARM), 160, 163, 177–78, 182–83, 184–85
International Comprehensive Ocean-Atmosphere Data Set (ICOADS), 96
International Council for the Exploration of the Sea (ICES), 169, 172
international development agencies, 159, 169–70. *See also* Asian Development Bank
International Rice Research Institute, 185

International Whaling Commission, 8, 133
Inuit, 149–50, 151, 262n509
IUU (illegal, unreported, and unregulated fisheries), 3

Jackson, Jeremy, 17, 165, 186
Japan: Alaska salmon fisheries and, 138–39; fishing-industrial complex, 22; international influence of whaling industry, 8, 133, 134–35, 136; seafood market, 10–11, 70, 165
jellyfish, 27
Johannes, R.E.: *Words of the Lagoon*, 38
Jones, Rodney, 172
judo arguments, 118

Kaschner, Kristin, 134–35
Kelvin, Lord, 110
Kiel University, 157, 170, 172, 185
Kittinger, J.N., 186
Kuhn, Thomas, 48, 102

Laevastu, T., 111
language, 113
Latin America, 36, 138–39
League of Nations, 1
Leeuwenhoek, Anton van, 116
Lenfest Ocean Program (LOP), 132
Lindeman, Raymond, 180
Lomborg, Bjorn, 13
Longhurst, A.R., 112
Ludwig, Don, 13

Magnuson-Stevens Fishery Conservation and Management Act, 68, 140, 141, 245n329
Malaysia, 73, 74, 76
Malthusian overfishing model: research on, 43–44; for small-scale fisheries, 39–41, 43–44 2004 tsunami example, 41–43;
Manila Bay (Philippines), 74
Maori, 50
Margalef, Ramon, 20
marine ecologists, 19, 25
marine fisheries. *See* fisheries

Marine Historical Ecology in Conservation (Kittinger, McClenachan, Gedan, and Blight), 186
marine reserves, 15, 19–20, 31–32, 57–58, 65–66, 144, 224n115, 224n119
Marine Stewardship Council (MSC): author's involvement with, 164, 187; consumer awareness focus, 11; problems with, 30, 166, 196, 217n35, 217n43, 227n142, 278n699
marine trophic index (MTI), 6, 217n38
Marisla Foundation, 184
market-based approaches, 62–63
Marr, Jack, 177, 178, 195
The Martian (film), 171
masking factors, 12–13, 119
mass-balance, 110–11
maximum sustainable yield (MSY), 95, 138–39, 140, 258n473
McClenachan, L., 186
memes, 48
Miller, Kristi, 129, 148
Misunderstanding Change (Rost), 187
Möbius, Karl, 157
models, 39, 102–3. *See also* Malthusian overfishing model
modern governance, 64
Montgomery, David, 162
Morissette, Lyne, 134–35
mortality, in fish, 172–73
Morton, Alexandra, 146–47
Mowat, Farley: *Sea of Slaughter*, 97
multidisciplinary work, 112
multispecies fisheries: individual transferable quotas and, 54; San Miguel Bay (Philippines), 178–79
Murphy, Garth, 3
Myanmar (Burma), 74
Myers, Ransom A., 17, 128, 165

National Oceanic and Atmospheric Administration (NOAA), 181, 182
Nauen, Cornelia, 183–84
Newfoundland and Labrador. *See* cod fishery, northern
The New Republic, 224n121

New Zealand, 50, 51–52, 235n225
North Sea, 4
Norway, 70, 129, 137
Nunavut, 150–51

Oak Foundation, 184
Obama, Barak H., 144
occurrence records (bioquads), 104–5
ocean acidification, 27–28, 226n139
oceanography, 96, 104
Odum, E.P., 180
oil sands, 129–30
Olivieri Affair, 127
omega-3 fatty acids, 10, 142
orange roughy fisheries, 21, 217n35, 217n43
otoliths, 176, 268n576
overfishing. *See* Malthusian overfishing model

Palau, 38
Palomares, Maria-Lourdes "Deng," 164, 180
paradigm shifts, 48, 102–3
parasites, 125, 126
Paul G. Allen Family Foundation, 87, 193
Pauly, Daniel: activism by, 163–64, 187; Arctic fisheries experience, 149–51; childhood, 156; citizenship, 155, 263n525; criticism of, 154–55; doctoral research program, 172; ecosystem modeling and Ecopath, 152–54, 180–82; education, 156–58, 159; ELEFAN program, 173–74; encounter with Department of Fisheries and Oceans, 148; engagement with Canadian history, 152; FishBase development, 182–84; fish growth theory, 174–77; future plans, 194–95; with ICLARM, 177–78, 184–85; in Indonesia, 158–59, 170–71; introduction to Canadian fisheries, 145–46; meta-analysis on natural mortality in fish, 172–73; Peruvian anchoveta fishery experience,

179–80; in Philippines, 160, 178–79, 185; potential Tanzanian job, 158, 170; racial background, 157; reflections on life, 195; salmon sea lice experience, 146–48; SeaLifeBase development, 184; at UBC, 185–86, 263n521; writings and contributions, 186–87. *See also* Sea Around Us
Pauly, Sandra Wade, 146
Pearse, Peter H., 52
pelagic realm, 5, 57
persistent organic pollutants, 10, 142
Peru: anchoveta fishery, 3, 169, 179–80
Pew Charitable Trusts, 87, 93, 164, 187–88, 193
Philippines: Bolinao reef fishery, 44; fish-aggregating devices and, 5; people-power revolution, 163; San Miguel Bay fisheries, 178–79; trawl industry, 73–74, 76
Pitcher, Tony, 185, 195
plate tectonics, 160–61
pollutants, persistent organic, 10, 142
pollution quotas, 49
Polovina, Jeffrey J., 111, 181
Ponzi schemes, 21–23, 225n122
population growth, of fish, 56, 236n239
privatization, 55–56. *See also* individual transferable quotas

quotas. *See* individual transferable quotas

red snapper, 56
reduction fisheries, 29, 30, 166
Rees, Bill, 161
refuges, natural, 57, 65. *See also* marine reserves
Regional Fishery Management Councils, 140, 258n476
regulation. *See* fisheries management
Reichert, Joshua, 187–88
remote sensing, 112
reserves. *See* marine reserves
Ricker, Bill, 3, 155

rights-based fishing, 18, 28–29, 45. *See also* individual transferable quotas
Romans, 77
Rost, D., 187
Rothschild, Brian, 177
Royal Canadian Mounted Police, 151, 262n509

Sala, E., 186
salmon: Alaska fisheries, 138–39; aquaculture issues, 129, 142, 146–48
salmon trees, 154
San Miguel Bay (Philippines), 178–79
sardine fishery (California), 2
Schaefer, Milner B., 139, 258n473
Schindler, David, 155
Schwann, Theodor, 116
science: approach to truth, 260n494; Canadian suppression of findings, 127–28, 128–29, 129–30; development process, 102–3; funding declines, 103–4; hierarchy of disciplines, 108; importance of appropriate focus, 116–17; judo arguments, 118; need for scientists to speak up, 130–31; standardization of key variables, 119–20. *See also* consilience
Sea Around Us: introduction, 222n96; catch data analysis project, 93, 164–65, 190; catch data database, 38, 87–88, 193–94, 233n216; catch reconstruction project, 14, 191–93; criticism of author's involvement, 154; establishment and funding, 164, 187–88, 193; *Global Atlas of Marine Fisheries*, 194, 243n316; mission and focus, 188–89, 190–91; progress benchmarks, 190; on small-scale fisheries, 71; stock-status plots and, 90–91
seafood: consumer awareness campaigns, 11, 30–31, 143–44, 166; ethical issues, 142–43; health issues, 142; U.S. relationship with, 141–42

Index

sea lice, 129, 146–47
SeaLifeBase, 87–88, 184, 246n338
Sea of Slaughter (Mowat), 97
Senegal, 44, 135, 136, 160
shame, 144
Shehadeh, Ziad, 178, 195
shifting baselines. *See* baselines, shifting
Shifting Baselines (Jackson, Alexander, and Sala), 186
shrimps, penaeid, 75–76
SimCoast software, 114, 250n383
skepticism, 13
small-scale fisheries: anthropologists research opportunities, 45–46; catch underestimation, 37–39, 71–72; competition with large-scale fisheries, 33; conflict in Southeast Asia with trawlers, 75, 76; economic and local importance, 26, 71–72; factors for decline, 33–34, 226n134; FAO voluntary guidelines, 233n216; fisheries economists research opportunities, 45; future possibilities, 47; individual transferable quotas and, 50–51; Malthusian overfishing model, 39–41, 43–44; return to in Southeast Asia, 78–79; sociologists research opportunities, 46; as sustainable, 242n306; traditional products from Southeast Asia, 76–77; 2004 tsunami example, 41–43
smoking, 100–101
social scientists, 34, 36–37, 38, 39, 43, 45–46, 228n154
Society of Canadian Limnologists, 131
sociologists, 46
Southeast Asia: bycatch by trawl industry, 75–76; conflict between small-scale fisheries and trawl industry, 75, 76; fish sauces, 76–78; history of trawl industry, 73–74, 76; return to small-scale fisheries, 78–79; traditional products from small-scale fisheries, 76–77; trash fish by trawl industry, 78

South Pacific, 8, 22, 37–39, 44, 71, 97
Spain, 5, 10, 216n23
Spam, 39, 231n191
Species 2000 Initiative, 104
Standard Social Science Model, 102
State of Fisheries and Aquaculture (FOA), 214n5
statistics. *See* catch data; catch reconstruction
statoliths, 176, 268n576
stock assessments, 51–52, 89, 92, 179
stock-status plots, 90–91, 194
Strait of Georgia, 152–53
sturgeon, Hudson River, 24
subsidies, 2, 11, 22, 31, 43, 70, 94, 106, 221n79
Sumaila, Rashid, 164
sustainability, 165, 168. *See also* consumer awareness campaigns; maximum sustainable yield
"Système d'Information et d'Analyse des Pêches de l'Afrique du Nord-Ouest" (SIAP), 133–34

tar sands, 129–30
taxonomic expansion, 6
taxonomy, 103
Thailand, 36, 41, 74
theories, 102, 246n332
thermodynamics, first law of, 110
Tiews, Klaus, 74
Tobin, Brian, 145–46
total allowable catch (TAC), 51, 95
trade, international, 10–11, 70–71, 142
tragedy of the commons, 3
transects, 114
trash fish, 77, 78
trawl industry: in Indonesia, 42, 158–59, 171; in Southeast Asia, 73–74, 75, 76; trash fish, 77, 78
Trinidad and Tobago, 85
Trudeau, Justin, 131
Trump, Donald, 144
Truth and Reconciliation Commission, 152
tsunami (2004), 41–43
tuna: aquaculture, 26–27, 143; fisheries decline, 23; health issues, 142;

maximum sustainable yield concept and, 138-39; South Pacific fisheries, 37-38, 39; traditional refuge of, 65
turbot war, 5, 216n23
2004 tsumani, 41-43

Ulanowicz, Robert, 181
uncertainty, scientific, 13
United Kingdom, 5, 9
United Nations Convention on the Law of the Sea (UNCLOS), 7, 69, 139, 227n144
United States of America: fish consumption and seafood imports, 12, 70, 165; fisheries history, 7, 137-38, 139-40; fisheries management, 32, 68, 140-41, 258n476; fishing-industrial complex, 22; individual transferable quotas and, 53; Magnuson-Stevens Act, 68, 140, 141, 245n329; maximum sustainable yield and, 138-39; relationship with seafood, 141-42; Trump and U.S. fisheries, 144; UNCLOS and, 139
universities, 127, 130, 131
University of British Columbia: author at, 185-86, 263n521; Fisheries Centre, 153, 154-55, 185, 186
University of Kiel, 157, 170, 172, 185
University of the Philippines, 185

variables, standardization of, 119-20
vessels, fishing, 9
Virtual Population Analysis, 180
voting rights, 101
Vulcan Inc., 87, 193

Wallace, Scott, 148, 261n498
wallet cards, 30-31, 143-44, 166
Walters, Carl J., 13, 52, 153, 182, 186
Watkinson, Steven, 153-54
Watson, Reg, 164

Watts, Paul, 150
Western Pacific Regional Fishery Management Council, 191-92
whales: baleen whales, 132-33; blamed for fisheries decline in Northwest Africa, 8, 132, 133, 134-35, 136; ecotourism, 136, 257n466; support for in Northwest Africa, 135-36
Wilson, E.O., 108
Wolff, Kasper Friedrich, 116
women, in small-scale fisheries, 46
Words of the Lagoon (Johannes), 38
World Bank, 26, 226n134
WorldFish. *See* International Center for Living Aquatic Resource Management (ICLARM)
World Trade Organization, 11
World Wide Fund for Nature (WWF), 132, 164, 187, 217n43

Yearbook of Fisheries and Aquaculture Statistics (FAO), 1-2, 89
yield-per-recruit analysis, 171

Zeller, Dirk, 164, 192

DAVID SUZUKI INSTITUTE

THE DAVID SUZUKI INSTITUTE is a non-profit organization founded in 2010 to stimulate debate and action on environmental issues. The Institute and the David Suzuki Foundation both work to advance awareness of environmental issues important to all Canadians.

We invite you to support the activities of the Institute. For more information please contact us at:

David Suzuki Institute
219 – 2211 West 4th Avenue
Vancouver, BC, Canada V6K 4S2
info@davidsuzukiinstitute.org
604-742-2899
www.davidsuzukiinstitute.org

Cheques can be made payable to The David Suzuki Institute.